SpringerBriefs in Physics

SpringerBriefs in Physics are a series of slim high-quality publications encompassing the entire spectrum of physics. Manuscripts for SpringerBriefs in Physics will be evaluated by Springer and by members of the Editorial Board. Proposals and other communication should be sent to your Publishing Editors at Springer.

Featuring compact volumes of 50 to 125 pages (approximately 20,000–45,000 words), Briefs are shorter than a conventional book but longer than a journal article. Thus Briefs serve as timely, concise tools for students, researchers, and professionals.

Typical texts for publication might include:

- A snapshot review of the current state of a hot or emerging field
- A concise introduction to core concepts that students must understand in order to make independent contributions
- An extended research report giving more details and discussion than is possible in a conventional journal article
- A manual describing underlying principles and best practices for an experimental technique
- An essay exploring new ideas within physics, related philosophical issues, or broader topics such as science and society

Briefs are characterized by fast, global electronic dissemination, straightforward publishing agreements, easy-to-use manuscript preparation and formatting guidelines, and expedited production schedules. We aim for publication 8–12 weeks after acceptance.

More information about this series at http://www.springer.com/series/8902

Silvia Viola Kusminskiy

Quantum Magnetism, Spin Waves, and Optical Cavities

 Springer

Silvia Viola Kusminskiy
Max-Planck-Institute for the Science
of Light
Erlangen, Bayern, Germany

ISSN 2191-5423   ISSN 2191-5431   (electronic)
SpringerBriefs in Physics
ISBN 978-3-030-13344-3   ISBN 978-3-030-13345-0   (eBook)
https://doi.org/10.1007/978-3-030-13345-0

Library of Congress Control Number: 2019931817

This Springer imprint is published by the registered company Springer Nature Switzerland AG
The registered company address is: Gewerbestrasse 11, 6330 Cham, Switzerland

Contents

Abstract

Both magnetic materials and light have always played a predominant role in information technologies, and continue to do so as we move into the realm of quantum technologies. In this course, we review the basics of magnetism and quantum mechanics, before going into more advanced subjects. Magnetism is intrinsically quantum mechanical in nature, and magnetic ordering can only be explained by the use of quantum theory. We will go over the interactions and the resulting Hamiltonian that governs magnetic phenomena, and discuss its elementary excitations, denominated magnons. After that we will study magneto-optical effects and derive the classical Faraday effect. We will then move on to the quantization of the electric field and the basics of optical cavities. This will allow us to understand a topic of current research denominated *Cavity Optomagnonics.*

This book is based on the notes written for the course I taught in the Summer Semester 2018 at the Friedrich-Alexander Universität in Erlangen. It is intended for Master or advanced Bachelor students. Basic knowledge of quantum mechanics, electromagnetism, and solid state at the bachelor level is assumed. Each section is followed by a couple of simple exercises which should serve as to "fill in the blanks" of what has been derived and a couple of checkpoints for the main concepts developed.

Chapter 1
Electromagnetism

The history of magnetism is ancient: just to give an example, the magnetic compass was invented in China more than 2000 years ago. The fact that magnetism is intrinsically connected to moving electric charges (and not to "magnetic charges"), however, was not discovered until much later. In the year 1820, Oersted experimentally demonstrated that a current-carrying wire had an effect on the orientation of a magnetic compass needle placed in its proximity. In the following few years, Ampere realized that a small current loop generates a magnetic field which is equivalent to that of a small magnet, and speculated that all magnetic fields are caused by charges in motion. In the next few sections, we will review these concepts and the basics of *magnetostatics*.

1.1 Basic Magnetostatics

As the name indicates, magnetostatics deals with magnetic fields that are constant in time. The condition for that is a steady-state current, in which both the charge density ρ and the current density $j = I/A_s$ (A_s cross-sectional area) are independent of time

$$\frac{\partial \rho}{\partial t} = 0 \tag{1.1.1}$$

$$\frac{\partial \mathbf{j}}{\partial t} = 0. \tag{1.1.2}$$

From the continuity equation

$$\nabla \cdot \mathbf{j} + \frac{\partial \rho}{\partial t} = 0, \tag{1.1.3}$$

© The Author(s), under exclusive license to Springer Nature Switzerland AG 2019
S. Viola Kusminskiy, *Quantum Magnetism, Spin Waves, and Optical Cavities*,
SpringerBriefs in Physics, https://doi.org/10.1007/978-3-030-13345-0_1

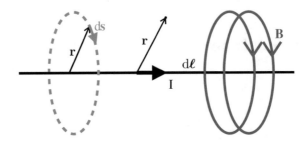

Fig. 1.1 The magnetic induction generated by a current I can be calculated using Biot–Savart's law, see Eq. 1.1.6. Ampere's law (see Eq. 1.1.7) is always valid, but useful to calculate the **B** fields only for cases of particular symmetry, e.g., an infinite straight wire

we moreover obtain

$$\nabla \cdot \mathbf{j} = 0. \tag{1.1.4}$$

In these notes, we will call *magnetic induction* to **B** and *magnetic field* to **H**.[1] In free space, these two fields are related by

$$\mathbf{B} = \mu_0 \mathbf{H} \tag{1.1.5}$$

being $\mu_0 = 4\pi \times 10^{-7} \text{NA}^{-2}$ the permeability of free space. We will use the SI units system throughout these notes, and therefore **B** is measured in Teslas $(\text{T} = \text{V.s.m}^{-2})$ and **H** in Amperes per meter (A.m^{-1}).

The magnetic induction at point **r** due to a current loop can be calculated using the Biot–Savart law

$$d\mathbf{B} = \frac{\mu_0 I}{4\pi r^2} d\boldsymbol{\ell} \times \frac{\mathbf{r}}{r}, \tag{1.1.6}$$

where $d\boldsymbol{\ell}$ points in the direction of the current I, see Fig. 1.1. Equivalent to the Biot–Savart law is Ampere's law, which reads

$$\oint_{\mathcal{C}} \mathbf{B} \cdot d\mathbf{s} = \mu_0 I, \tag{1.1.7}$$

where I is the current enclosed by the closed loop $\mathcal{C}$, see Fig. 1.1. Ampere's law is general, but it is useful to *calculate* magnetic fields only in cases of high symmetry, for example, the magnetic field generated by an infinite straight wire. Using Stoke's theorem, we can put Ampere's law in differential form

$$\nabla \times \mathbf{B} = \mu_0 \mathbf{j}. \tag{1.1.8}$$

---

[1]Some authors call instead **B** the magnetic field and **H** the *auxiliary field*.

Ampere's law together with the absence of magnetic monopoles condition

$$\nabla \cdot \mathbf{B} = 0 \tag{1.1.9}$$

constitute the *Maxwell equations for magnetostatics*. These equations give us indeed time-independent magnetic fields, and if we compare the magnetostatic equations with the full microscopic Maxwell equations

$$\nabla \cdot \mathbf{B} = 0 \tag{1.1.10}$$

$$\nabla \cdot \mathbf{E} = \frac{\rho}{\varepsilon_0} \tag{1.1.11}$$

$$\nabla \times \mathbf{E} = -\frac{\partial \mathbf{B}}{\partial t} \tag{1.1.12}$$

$$\nabla \times \mathbf{B} = \mu_0 \left(\mathbf{j} + \varepsilon_0 \frac{\partial \mathbf{E}}{\partial t} \right) \tag{1.1.13}$$

(with $\varepsilon_0 = 8.85 \times 10^{-12}\,\mathrm{Fm}^{-1}$ the vacuum permittivity) we see that we have, moreover, decoupled the magnetic and electric fields.

Check Points

- What is the magnetostatic condition?
- Write the magnetostatic Maxwell equations.

1.2 Magnetic Moment

The magnetic moment of a current loop is defined as

$$\mathbf{m} = I A \hat{\mathbf{n}}, \tag{1.2.1}$$

where A is the area enclosed by the loop and $\hat{\mathbf{n}}$ is the normal to the surface, with its direction defined from the circulating current by the right-hand rule, see Fig. 1.2. $\mathbf{m}$ defines a *magnetic dipole* in the limit of $A \to 0$ but finite moment.

Using Eq. 1.1.6, we can calculate the magnetic induction generated by a small current loop of radius R

$$\mathbf{B}(\mathbf{r}) = \frac{\mu_0 I}{4\pi} \int \frac{d\boldsymbol{\ell}' \times \Delta\mathbf{r}}{\Delta r^3} \tag{1.2.2}$$

$$= -\frac{\mu_0 I}{4\pi} \int d\boldsymbol{\ell}' \times \nabla \left(\frac{1}{\Delta r} \right)$$

with $\Delta\mathbf{r} = \mathbf{r} - \mathbf{r}'$ (see Fig. 1.3). From Eq. 1.1.9, we know we can define a vector potential $\mathbf{A}(\mathbf{r})$ such that $\mathbf{B}(\mathbf{r}) = \nabla \times \mathbf{A}(\mathbf{r})$. By a simple manipulation of Eq. 1.2.2, one can show that in the far-field limit ($\Delta\mathbf{r} \gg R$),

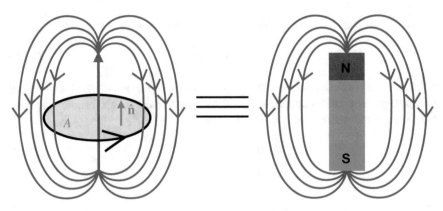

Fig. 1.2 Magnetic dipole: the magnetic field induced by a small current loop is equivalent to that of a small magnet

Fig. 1.3 Magnetic induction due to a small circular current loop: we use Biot–Savart to calculate the **B** field. Seen from "far away", it is the field of a magnetic dipole

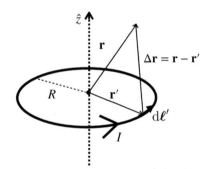

$$\mathbf{A}(\mathbf{r}) = \frac{\mu_0}{4\pi} \mathbf{m} \times \frac{\hat{\mathbf{r}}}{r^2} \tag{1.2.3}$$

$$\mathbf{B}(\mathbf{r}) = \frac{\mu_0}{4\pi} \frac{3(\mathbf{m} \cdot \mathbf{r})\mathbf{r} - r^2\mathbf{m}}{r^5}, \tag{1.2.4}$$

which is the magnetic induction generated by a magnetic dipole. More generally, for an arbitrary current density distribution $\mathbf{j}(\mathbf{r}')$, one can define [1, 2]

$$\mathbf{m} = \frac{1}{2} \int d^3\mathbf{r}' \left[\mathbf{r}' \times \mathbf{j}(\mathbf{r}')\right] \tag{1.2.5}$$

and Eq. 1.2.3 is the lowest nonvanishing term in a multipole expansion of the vector potential (in the Coulomb gauge, $\nabla \cdot \mathbf{A} = 0$)

$$\mathbf{A}(\mathbf{r}) = \frac{\mu_0}{4\pi} \int d^3\mathbf{r}' \frac{\mathbf{j}(\mathbf{r}')}{|\mathbf{r} - \mathbf{r}'|}. \tag{1.2.6}$$

The energy of a magnetic dipole in a magnetic field is given by

$$E_Z = -\mathbf{m} \cdot \mathbf{B} \tag{1.2.7}$$

and therefore is minimized for $\mathbf{m} \parallel \mathbf{B}$. This is called the *Zeeman Energy*.

1. *Exercise: derive Eqs.* 1.2.3 *and* 1.2.4 *(tip: use the "chain rule" and a multipole expansion).*
2. *Exercise: show that Eq.* 1.2.1 *follows from* 1.2.5 *(tip: 1-D Delta-function distributions have units of 1/length).*

Check Points

- How do you show the equivalence between the magnetic field of a small current loop and that of a small magnet? A conceptual explanation suffices.

1.3   Orbital Angular Momentum

The magnetic moment $\mathbf{m}$ can be related to angular momentum. In order to do this, we consider the limit of one electron e (with negative charge $-e$) orbiting around a fixed nucleus, see Fig. 1.4. Note that here we get the first indication that magnetism is a purely quantum effect: stable orbits like that are not allowed classically, and we need quantum mechanics to justify the stability of atoms. Our argument here is therefore a semiclassical one. The average current due to this single electron is

$$I = -\frac{e}{T} = -\frac{e\omega}{2\pi}, \tag{1.3.1}$$

where T is one period of revolution. The electron also possesses orbital angular momentum $\mathbf{L} = m_e \mathbf{r} \times \mathbf{v}$. Measured from the center of the orbit,

$$\mathbf{L} = m_e R^2 \vec{\omega} \tag{1.3.2}$$

and using Eq. 1.2.1, we obtain

$$\mathbf{m} = -\frac{e}{2m_e}\mathbf{L}. \tag{1.3.3}$$

Therefore, we have linked the magnetic moment of a moving charge to its orbital angular momentum. The coefficient of proportionality is called the *gyromagnetic ratio*

$$\gamma_L = -\frac{e}{2m_e}, \tag{1.3.4}$$

which is negative due to the negative charge of the electron. Hence, in this case the magnetic moment and angular momentum are antiparallel. In solids, electrons are the primary source of magnetism due to their small mass compared to that of

Fig. 1.4 Semiclassical picture used to calculate the orbital angular momentum of an electron

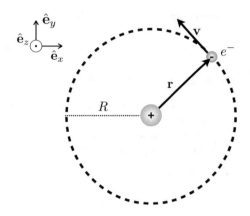

the nucleus. Since $m_p \approx 10^3 m_e$, the gyromagnetic ratio for the nucleus is strongly suppressed with respect to the electronic one.

Check Points

- What is the gyromagnetic ratio?
- Why is the gyromagnetic ratio of the nucleus suppressed with respect to the electronic one?

1.4 Spin Angular Momentum

Although we performed a classical calculation, the result obtained for the gyromagnetic ratio in Eq. 1.3.4 is consistent with the quantum mechanical result. We know, however, that the electron posses an intrinsic angular momentum, that is, the *spin* **S**. The *total* angular momentum of the electron is therefore given by

$$\mathbf{J} = \mathbf{L} + \mathbf{S} . \tag{1.4.1}$$

The spin has no classical analog and the coefficient of proportionality γ_S between magnetic moment and spin

$$\mathbf{m}_S = \gamma_S \mathbf{S} \tag{1.4.2}$$

needs to be calculated quantum mechanically *via* the Dirac equation (see, e.g., Chap. 2 of Ref. [3]). The result is

$$\gamma_S \approx -\frac{e}{m} = 2\gamma_L , \tag{1.4.3}$$

where the approximate symbol indicates that there are relativistic corrections (also contained in the Dirac equation!) to this expression. The γ_S value agrees with experimental observations.

Fig. 1.5 A magnetic
moment in a **B** field
experiences a torque
$\mathcal{T} = \mathbf{m} \times \mathbf{B}$. Remember: for
an electron, **L** and **m** point in
opposite directions

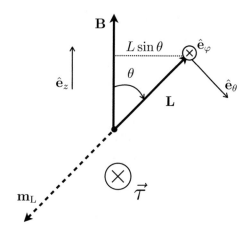

The total magnetic moment of the electron is therefore given by

$$\mathbf{m}_{\text{TOT}} \approx \gamma_L \left(\mathbf{L} + 2\mathbf{S} \right), \tag{1.4.4}$$

and hence is not simply proportional to the total angular momentum! To understand
the relation between $\mathbf{m}_{\text{TOT}}$ and **J**, given by the *Landé factor*, we need to resort to
quantum mechanics and the operator representation of angular momentum. We will
do that in the next chapter, where we discuss the atomic origins of magnetism.

Check Points

- What is the relation between the magnetic moment of an electron and the angular
  momentum operators?

1.5 Magnetic Moment in a Magnetic Field

A magnetic moment in a magnetic field experiences a torque

$$\mathcal{T} = \mathbf{m} \times \mathbf{B}. \tag{1.5.1}$$

Therefore, the classical equation of motion for the magnetic dipole (considering for
the moment only the orbital angular momentum) is

$$\frac{d\mathbf{L}}{dt} = \mathbf{m} \times \mathbf{B} = \gamma_L \mathbf{L} \times \mathbf{B}. \tag{1.5.2}$$

Using the geometry depicted in Fig. 1.6, we obtain

Fig. 1.6 Coordinates used for solving the equation of motion Eq. 1.5.2 viewed "from above", in a plane perpendicular to **B**

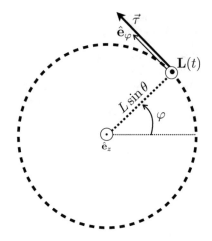

$$\dot{\mathbf{L}} = L \sin \theta \dot{\varphi} \mathbf{e}_\phi \qquad (1.5.3)$$

$$\gamma_L \mathbf{L} \times \mathbf{B} = |\gamma_L| L B \sin \theta \mathbf{e}_\phi . \qquad (1.5.4)$$

Hence, the magnetic moment precesses around **B** at a frequency

$$\omega_L = \dot{\varphi} = |\gamma_L \mathbf{B}|, \qquad (1.5.5)$$

which is denominated the *Larmor frequency*. Therefore, the angular momentum will precess around the **B** field at a fixed angle θ and with constant angular frequency ω_L.

This is consistent with the energy expression defined in Eq. 1.2.7. The work per unit time performed by the torque is given by the usual expression for the power

$$\frac{dW}{dt} = \mathcal{T} \cdot \vec{\omega}, \qquad (1.5.6)$$

where the angular velocity vector is perpendicular to the plane of rotation and its direction is given by the right-hand rule. We see therefore that there is no power transfer in the Larmor precession, since $\omega_L \hat{\mathbf{e}}_z \cdot \mathcal{T} = 0$. There is, however, an energy cost if we want to change the angle θ of precession, since the resultant angular velocity $\dot{\theta} \mathbf{e}_\varphi$ is collinear with the torque. Using (for simplicity we defined now θ as the angle between **m** and **B**, see Fig. 1.7)

$$\mathcal{T} = -mB \sin \theta \mathbf{e}_\varphi$$

$$\vec{\omega}_\theta = -\dot{\theta} \mathbf{e}_\varphi, \qquad (1.5.7)$$

we find

$$\mathcal{T} \dot{\theta} = -mB \sin \theta \frac{d\theta}{dt} \qquad (1.5.8)$$

Fig. 1.7 Coordinate system
used to obtain Eq. 1.5.9

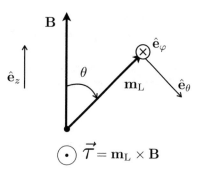

and therefore the work exerted to rotate **m** up to an angle θ is (up to a constant)

$$W = -\int mB \sin\theta d\theta = \mathbf{m} \cdot \mathbf{B} = -E_Z. \tag{1.5.9}$$

We can therefore take the Zeeman energy E_Z as the potential energy associated with the necessary work required to rotate the dipole **m** with respect to an external **B** field.

We will see equations of motion in the form of Eq. 1.5.2 reappearing throughout this course, even as we treat the angular momenta as quantum operators. The reason is that, even though the total magnetic moment $\mathbf{m}_{TOT}$ is not proportional to the total angular momentum **J** (see Eq. 1.4.4), their quantum mechanical expectation values are proportional to each other through the Landé factor g. We will see this more formally when we start dealing with the quantum mechanical representation of the angular momenta. For now, we assume that $\mathbf{m}_{TOT}$ and **J** are related by

$$\mathbf{m}_{TOT} \cdot \mathbf{J} = g\gamma_L \mathbf{J} \cdot \mathbf{J} \tag{1.5.10}$$

from which we can obtain a classical expression for g, by replacing Eqs. 1.4.4 and 1.4.1 into 1.5.10 and noting that

$$\mathbf{L} \cdot \mathbf{S} = \frac{1}{2} \left(J^2 - L^2 - S^2 \right)$$

from $J^2 = (\mathbf{L} + \mathbf{S})^2$. One obtains

$$g_{cl} = \frac{3}{2} + \frac{S^2 - L^2}{2J^2}, \tag{1.5.11}$$

where the superscript indicates this is a classical approximation for g, which coincides with the quantum mechanical result in the limit J^2, S^2, and L^2 large [4].

Check Points

- Write the equation of motion for an angular momentum in the presence of a magnetic field.
- What is the dynamics of an angular momentum in the presence of a magnetic field?
- What is the Larmor frequency?

1.6 Magnetization

Inside a material, the magnetic induction **B** indicates the response of the material to the applied magnetic field **H**. Both vector fields are related through the *magnetization* in the sample

$$\mathbf{B} = \mu_0 \left(\mathbf{H} + \mathbf{M} \right) , \tag{1.6.1}$$

where the magnetization is defined as the average magnetic moment per unit volume,

$$\mathbf{M}(\mathbf{r}) = \frac{\langle \mathbf{m} \rangle_V}{V} , \tag{1.6.2}$$

and where the average indicates that we average over all atomic magnetic moments in a small volume V around position $\mathbf{r}$.[2] In this way, a smooth vectorial function of position is obtained. From Eq. 1.6.1, we see that the magnetization has the same units as the magnetic field **H** ($A.m^{-1}$). In Eq. 1.6.1, both **B** and **H** indicate the fields *inside* the material, and hence **H** contains also the *demagnetizing* fields (that is, it is not just the external applied field). We will see more on demagnetization fields in the next section.

The response to the magnetic field of the magnetization and field induction are characterized by the *magnetic susceptibility* χ and the *permeability* μ, respectively

$$\mathbf{M} = \chi \mathbf{H} \tag{1.6.3}$$

$$\mathbf{B} = \mu \mathbf{H} , \tag{1.6.4}$$

where we have written the simplest expressions for the case in which all fields are collinear, static (that is, independent of time), and homogeneous in space ($\mathbf{q} = \omega = 0$). In general, however, the response functions are tensorial quantities, e.g., $M_i = \sum_j \chi_{ij} H_j$, and depend on frequency ω and momentum $\mathbf{q}$. Note that from Eq. 1.6.1, we obtain

$$\mu_{\mathrm{r}} = \frac{\mu}{\mu_0} = 1 + \chi \tag{1.6.5}$$

again in the simple collinear case. μ_{r} is the *relative permeability*, is dimensionless, and equals to unity in free space.

---

[2] We average over a "microscopically large but macroscopically small" volume V.

The quantities defined in Eqs. 1.6.3 and 1.6.4 are still allowed to depend on temperature T and magnetic field $\mathbf{H}$. We will now consider qualitatively the dependence on $\mathbf{H}$. For linear materials, χ and μ are independent of $\mathbf{H}$. A linear material with negative constant susceptibility is *diamagnetic,* whereas a positive susceptibility indicates either *paramagnetism* (no magnetic order) or *antiferromagnetism* (magnetic order with magnetic moments anti-aligned and zero total magnetization). In these cases, the magnetization is finite only in the presence of a magnetic field. On the other hand, if χ and μ depend on $\mathbf{H}$, the relations Eqs. 1.6.3 and 1.6.4 are nonlinear. This is the case for magnetically ordered states with net magnetization, namely, *ferromagnets* (magnetic moments aligned and pointing in the same direction) and *ferrimagnets* (magnetic moments anti-aligned but of different magnitudes, so that there is a net magnetization). In these materials, the magnetization increases nonlinearly with the applied field and saturates when all the magnetic moments are aligned. When decreasing the magnetic field, there is a remanent, finite magnetization at zero field. This process is called *hysteresis* and it is used to magnetize materials. As we learned in the previous section, the magnetic moment, and hence the magnetic characteristics of a material, are related to the total angular momentum of the electrons, and therefore on the atomic structure. We will learn more about this in the next chapter.

Check Points

- What is the relation between magnetic moment and magnetization?

1.7 Magnetostatic Maxwell Equations in Matter

To calculate the magnetic dipole moment $\mathbf{m}$ from Eq. 1.2.5, we have to know the microscopic current density. In general, however, we are not interested in microscopic, fast fluctuations. We already saw an example in which we considered the average current I generated by one orbiting electron, to obtain semiclassically the gyromagnetic ratio γ_L in Sec. 1.3. We have also defined the magnetization $\mathbf{M}$ as a macroscopic quantity which entails the average density of the microscopic $\mathbf{m}$. In a material, in general, we have access to the magnetization, which is due to bound microscopic currents, and to the macroscopic current density due to free charges, which we will denominate $\mathbf{j}_F$. This motivates defining a macroscopic vector potential $\mathbf{A}$ in terms of these two macroscopic quantities, and not the microscopic currents as in Eq. 1.2.6

$$\mathbf{A}(\mathbf{r}) = \frac{\mu_0}{4\pi} \int d^3 r' \left[\frac{\mathbf{j}_F(\mathbf{r}')}{|\mathbf{r} - \mathbf{r}'|} + \frac{\mathbf{M}(\mathbf{r}') \times (\mathbf{r} - \mathbf{r}')}{|\mathbf{r} - \mathbf{r}'|^3} \right]. \tag{1.7.1}$$

Note that this is simply rewriting Eq. 1.2.6, separating the bound- and free-current contributions. The bound-current contribution, the second term in Eq. 1.7.1, is written in terms of the magnetization and is equivalent to an averaged Eq. 1.2.3.

Equation 1.7.1 allows us to define an effective current density associated with the magnetization, by noting that [1]

$$\int_{\mathcal{V}} d^3r' \frac{\mathbf{M}(\mathbf{r}') \times (\mathbf{r} - \mathbf{r}')}{|\mathbf{r} - \mathbf{r}'|^3} = \int_{\mathcal{V}} d^3r' \mathbf{M}(\mathbf{r}') \times \nabla' \left(\frac{1}{|\mathbf{r} - \mathbf{r}'|} \right) \tag{1.7.2}$$

$$= \int_{\mathcal{V}} d^3r' \nabla' \times \mathbf{M}(\mathbf{r}') \left(\frac{1}{|\mathbf{r} - \mathbf{r}'|} \right) + \oint_{\mathcal{S}} \frac{\mathbf{M}(\mathbf{r}') \times d\mathbf{a}'}{|\mathbf{r} - \mathbf{r}'|}.$$

We can therefore define an effective *bound volume current density*

$$\mathbf{j}_B = \nabla \times \mathbf{M} \tag{1.7.3}$$

and an effective *bound surface current density*

$$\mathbf{K}_B = \mathbf{M} \times \hat{\mathbf{n}}, \tag{1.7.4}$$

where the surface element is defined as $d\mathbf{a} = d a \hat{\mathbf{n}}$. In the bulk, for a well-behaved magnetization function, the surface integral vanishes and we obtain

$$\mathbf{A}(\mathbf{r}) = \frac{\mu_0}{4\pi} \int d^3r' \left[\frac{\mathbf{j}_F(\mathbf{r}')}{|\mathbf{r} - \mathbf{r}'|} + \frac{\mathbf{j}_B(\mathbf{r}')}{|\mathbf{r} - \mathbf{r}'|} \right]. \tag{1.7.5}$$

The surface current $\mathbf{K}_B$ enters usually through boundary conditions at interfaces.

If we now go back to Ampere's Eq. 1.1.8 and separate the total current density into free and bound contributions $\mathbf{j} = \mathbf{j}_F + \mathbf{j}_B$, we obtain

$$\nabla \times \mathbf{B} = \mu_0 \left(\mathbf{j}_F + \nabla \times \mathbf{M} \right) \tag{1.7.6}$$

which defines the magnetic field $\mathbf{H}$

$$\mathbf{H} = \frac{1}{\mu_0} \mathbf{B} - \mathbf{M} \tag{1.7.7}$$

such that

$$\nabla \times \mathbf{H} = \mathbf{j}_F. \tag{1.7.8}$$

Therefore, the magnetic field $\mathbf{H}$ takes into account in an average way the bound currents, and has as its only source the free currents. Equation 1.7.8 is equivalent to Eq. 1.1.8, just rewritten in a more convenient form for macroscopic magnetostatics in matter. Note that $\mathbf{H}$ is, on the contrary to $\mathbf{B}$, not divergence-free:

$$\nabla \cdot \mathbf{H} = -\nabla \cdot \mathbf{M}. \tag{1.7.9}$$

The magnetostatic Maxwell equations in matter (also known as "macroscopic") therefore read

$$\nabla \cdot \mathbf{B} = 0$$
$$\nabla \times \mathbf{H} = \mathbf{j}_F \qquad (1.7.10)$$

and have to be complemented by the *constitutive equation* $\mathbf{B} = \mu\mathbf{H}$ in linear media (taking μ as a constant, independent of $\mathbf{H}$) or $\mathbf{B} = F(\mathbf{H})$ in nonlinear (e.g., ferromagnetic) media, where F is a characteristic function of the material.

We finish this section by stating the magnetostatic boundary conditions at an interface between two different media 1 and 2

$$(\mathbf{B}_2 - \mathbf{B}_1) \cdot \hat{\mathbf{n}} = 0 \qquad (1.7.11)$$
$$\hat{\mathbf{n}} \times (\mathbf{H}_2 - \mathbf{H}_1) = \mathbf{K}_F, \qquad (1.7.12)$$

where $\mathbf{K}_F$ is a free surface current density (usually 0).

1. *Exercise: Prove Eq.* 1.7.2.

Check Points

- What is the meaning of the magnetic field $\mathbf{H}$?
- What are the magnetostatic Maxwell equations in matter?

1.8 Demagnetizing Fields

A crucial difference between magnetic and electric fields is the lack of free magnetic charges or monopoles.[3] They are, however, a useful mathematical construction in some cases, for example, to calculate the so-called demagnetization fields. In finite systems, we can consider the magnetization as dropping to zero abruptly at the boundary of the material, giving rise to an accumulated "magnetic charge density" at the surface which acts as an extra source of magnetic fields inside of the material. These fields in general oppose to an externally applied magnetic field and are therefore dubbed *demagnetizing* fields. A surface magnetic charge density is energetically costly, and for finite magnetic systems at the microscale, it can determine the spatial dependence of the magnetically ordered ground state, giving rise to *magnetic textures*.

If we consider the special case of no free currents, $\mathbf{j}_F = 0$, Eqs. 1.7.10 imply that we can define a magnetic scalar potential ϕ_M such that

$$\mathbf{H} = -\nabla\phi_M \qquad (1.8.1)$$

---

[3]Magnetic monopoles, if they exist, have evaded experimental detection so far. They can, however, emerge as effective quasiparticles in condensed matter systems, and have been detected in materials which behave magnetically as a "spin ice" [5–7].

Fig. 1.8 Coordinate system for the uniformly magnetized sphere problem

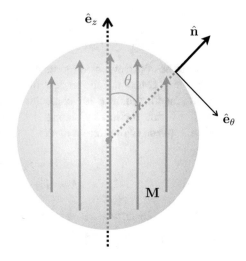

and using Eq. 1.7.9 we obtain a Poisson equation

$$\nabla^2 \phi_M = -\nabla \cdot \mathbf{M} \tag{1.8.2}$$

with solution

$$\phi_M(\mathbf{r}) = -\frac{1}{4\pi} \int_{\mathcal{V}} d^3 r' \frac{\nabla \cdot \mathbf{M}(\mathbf{r}')}{|\mathbf{r} - \mathbf{r}'|} + \frac{1}{4\pi} \oint_{\mathcal{S}} da' \frac{\hat{\mathbf{n}}' \cdot \mathbf{M}(\mathbf{r}')}{|\mathbf{r} - \mathbf{r}'|} . \tag{1.8.3}$$

Analogous to the case of the vector potential in Eq. 1.7.2, this allows us to define an effective *magnetic charge density*

$$\rho_M = -\nabla \cdot \mathbf{M} \tag{1.8.4}$$

and an effective *magnetic surface-charge density*

$$\sigma_M = \mathbf{M} \cdot \hat{\mathbf{n}} . \tag{1.8.5}$$

We see that ρ_M can only be finite for a nonhomogeneous magnetization $\mathbf{M}(\mathbf{r})$, whereas a finite σ_M indicates a discontinuity of $\mathbf{M}$ at the chosen surface $\mathcal{S}$.

1. *Exercise: Uniformly magnetized sphere*

For a ferromagnet at saturation, the magnetization can be considered as given, so we can in principle calculate the resulting magnetic field for a given geometry using Eq. 1.8.3. We consider here as an example the case of a uniformly magnetized sphere as depicted in Fig. 1.8.

(a) Choosing $\hat{\mathbf{e}}_z$ in the direction of $\mathbf{M}$, we can write $\mathbf{M} = M_0 \hat{\mathbf{e}}_z$. Calculate ρ_M and σ_M and write the Poisson equation for ϕ_M.

(b) Show that the scalar potential inside of the sphere is

$$\phi_M^{in} = \frac{1}{3} M_0 z \qquad (1.8.6)$$

and find the magnetic field $\mathbf{H}^{in}$ and magnetic induction $\mathbf{B}^{in}$ inside of the sphere.

(c) The magnetic field $\mathbf{H}^{in}$ inside of the sphere opposes the magnetization and it is therefore called a *demagnetizing field*. The proportionality coefficient between $\mathbf{H}^{in}$ and $\mathbf{M}$ is called the *demagnetizing factor N*. What is the value of N in this case? Demagnetization factors are geometry dependent and can moreover be defined only in very special cases with simple geometries.[4] Besides the sphere, one can define demagnetization factors for an infinite plane, an infinite cylinder, and a spheroid.

(d) Let us assume that now the sphere is placed in an external magnetic field $\mathbf{H}_0$. Using linearity, write the solution for $\mathbf{H}^{in}$ and $\mathbf{B}^{in}$ in this case.

(e) Let us now consider the case that the sphere is not permanently magnetized, but we now the material has a permeability μ. From the constitutive equation

$$\mathbf{B}^{in} = \mu \mathbf{H}^{in}, \qquad (1.8.7)$$

obtain the magnetization as a function of the external magnetic field $\mathbf{M}_\mu(\mathbf{H}_0)$, where the notation $\mathbf{M}_\mu$ implies that in this case we consider the magnetization not as given, but it depends on the permeability of the material. Show that $\mathbf{M}_\mu(0) = 0$, and therefore the obtained expression is not valid for materials with permanent magnetization.

Check Points

• Which is the origin of the demagnetization factors?

---

[4]The demagnetizing fields are always present, but it is only in very simple geometries that one can describe them with simple numerical factors.

Chapter 2
Atomic Origins of Magnetism

In the previous chapter, we reviewed the basic concepts of magnetism and magnetostatics using some semiclassical considerations. In particular, we attributed the magnetic moment of atoms to "small current loops" and to the angular momentum of electrons. In this chapter, we will put these concepts into more solid footing with the help of quantum mechanics.

2.1 Basics of Quantum Mechanics

We first review some basic concepts of quantum mechanics. In quantum mechanics, we describe a particle of mass m in a potential V by a wavefunction $\psi(\mathbf{r}, t)$ which satisfies the *Schrödinger equation*

$$i\hbar \frac{\partial \psi(\mathbf{r}, t)}{\partial t} = -\frac{\hbar^2}{2m} \nabla^2 \psi(\mathbf{r}, t) + V \, \psi(\mathbf{r}, t) . \tag{2.1.1}$$

The probability of finding the particle at a time t in a volume element $\mathrm{d}^3 r$ around position $\mathbf{r}$ is given by $|\psi(\mathbf{r}, t)|^2 \mathrm{d}^3 r$. If the potential V is independent of time, $\psi(\mathbf{r}, t) = \psi(\mathbf{r}) f(t)$ and $\psi(\mathbf{r})$ is an eigenfunction of the *time-independent Schrödinger equation*

$$-\frac{\hbar^2}{2m} \nabla^2 \psi(\mathbf{r}) + V(\mathbf{r}) \, \psi(\mathbf{r}) = E \psi(\mathbf{r}) \tag{2.1.2}$$

with energy E. Equivalently, we can write the eigenvalue equation for the Hamiltonian in Dirac notation

$$\hat{H} |\psi\rangle = E |\psi\rangle, \tag{2.1.3}$$

© The Author(s), under exclusive license to Springer Nature Switzerland AG 2019
S. Viola Kusminskiy, *Quantum Magnetism, Spin Waves, and Optical Cavities*,
SpringerBriefs in Physics, https://doi.org/10.1007/978-3-030-13345-0_2

where the quantum state of the particle is represented by the ket $|\psi\rangle$, the respective wavefunction is $\psi(\mathbf{r}) = \langle \mathbf{r}|\psi\rangle$, and the Hamiltonian operator is

$$\hat{H} = \frac{\hat{p}^2}{2m} + V(\hat{\mathbf{r}}). \tag{2.1.4}$$

In the position representation, $\hat{\mathbf{p}} \rightarrow -i\hbar\nabla$ so that, for example, $\langle \mathbf{r}|\hat{\mathbf{p}}|\psi\rangle = -i\hbar\nabla\psi(\mathbf{r})$.

In general, for any operator $\hat{A}$ we can write the *eigenvalue equation* $\hat{A}|\psi_\alpha\rangle = \alpha|\psi_\alpha\rangle$, where $|\psi_\alpha\rangle$ is an eigenstate with eigenvalue α. For a Hermitian operator, $\hat{A}^\dagger = \hat{A}$, α is real and the eigenstates form a basis of the Hilbert space where the operator acts. This is called an *observable*. The expectation value for $\hat{A}$ if the system is in the eigenstate $|\psi_\alpha\rangle$ then is simply $\langle \psi_\alpha \hat{A}|\psi_\alpha\rangle = \alpha$. If we consider a second operator $\hat{B}$ acting on the same Hilbert space, it is only possible to find a common basis of eigenstates of $\hat{A}$ and $\hat{B}$ if and only if the two operators commute: $[\hat{A}, \hat{B}] = \hat{A}\hat{B} - \hat{B}\hat{A} = 0$. In this case, the two operators can be measured simultaneously to (in principle) arbitrary precision. If the operators do not commute, then we run into the *Heisenberg uncertainty principle*. The most well-known example is that of the momentum and position operators, which satisfy $[\hat{x}, \hat{p}] = i\hbar$. How precise we measure one of the operators will determine the precision up to which we can know the value of the other: $\Delta x \Delta p \geq \hbar/2$. In general,

$$\Delta A \Delta B \geq \frac{1}{2}\left|\langle [\hat{A}, \hat{B}]\rangle\right|, \tag{2.1.5}$$

where $\Delta A = \sqrt{\langle \hat{A}^2\rangle - \langle \hat{A}\rangle^2}$ corresponds to the standard variation of $\hat{A}$ and analogously for operator $\hat{B}$.

2.2 Orbital Angular Momentum in Quantum Mechanics

The orbital angular momentum operator expression in quantum mechanics is inherited from its classical expression, $\hat{\mathbf{L}} = \hat{\mathbf{r}} \times \hat{\mathbf{p}}$. In the position representation, it is given by

$$\hat{\mathbf{L}} = -i\hbar\mathbf{r} \times \nabla. \tag{2.2.1}$$

From this expression, it is easy to verify that the different components of $\hat{\mathbf{L}}$ do not commute with each other. Instead, one obtains

$$[\hat{L}_i, \hat{L}_j] = i\hbar\epsilon_{ijk}\hat{L}_k, \tag{2.2.2}$$

where ϵ_{ijk} is the Levi-Civita tensor and the Einstein convention for the implicit sum or repeated indices has been used. Therefore, it is not possible to measure simultaneously with arbitrary precision all components of the angular momentum.

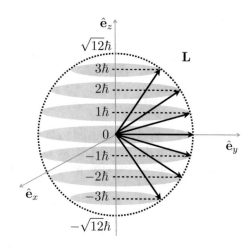

Fig. 2.1 Pictorial representation of the spatial quantization of angular momentum. In this example, $l = 3$. **L** precesses around the z-axis and has a definite projection on it that can take one of the allowed values $-3 \leq m_l \leq 3$. Note that the maximal quantum mechanically allowed value of the projection ($3\hbar$ in this case) is smaller than the classically allowed one

Let us assume we choose to measure $\hat{L}_z$. In this case, the Heisenberg uncertainty principle reads

$$\Delta L_x \Delta L_y \geq \frac{\hbar}{2} |\langle L_z \rangle| . \qquad (2.2.3)$$

It is, however, possible to find a common basis for $\hat{L}^2$ and one of the angular momentum components, since $[\hat{L}^2, \hat{L}_i] = 0$. Typically, $\hat{L}_z$ is taken and the respective eigenvalues are labeled by l and m. These are called the *quantum numbers*. The eigenstates satisfy

$$\hat{L}^2 |\psi_{lm_l}\rangle = \hbar^2 l(l+1)|\psi_{lm_l}\rangle$$
$$\hat{L}_z |\psi_{lm_l}\rangle = \hbar m |\psi_{lml}\rangle . \qquad (2.2.4)$$

For the orbital angular momentum, l is an integer and $-l \leq m_l \leq l$. These conditions can be depicted pictorially as in Fig. 2.1. The angular momentum vector has a magnitude $\hbar\sqrt{l(l+1)}$, and its projection on the z-axis is quantized and takes one of the possible values $\hbar m_l$. The maximum value of L_z is $\hbar l$, instead of $\hbar\sqrt{l(l+1)}$ as one would expect classically. We recover the classical expectation in the limit $l \gg 1$. The L_x and L_y components do not have a definite value and are represented as a precession of **L** around the z-axis.

Check Points

- Explain graphically the properties of an angular momentum operator and how it differs from a classical angular momentum.

2.3 Hydrogen Atom

For a problem with rotational symmetry, the angular momentum is conserved: $[\hat{H}, \hat{\mathbf{L}}] = 0$. Hence, we can find a basis of eigenstates that are also energy eigenstates, such that

$$\hat{H}|\psi_{nlm}\rangle = E_n|\psi_{nlm}\rangle. \tag{2.3.1}$$

The quantum numbers are denominated as *principal*, *azimuthal*, and *magnetic*, respectively, for n, l, and m. If we ignore spin, we have all the tools to solve the energy levels and orbitals for the hydrogen atom, in which the electron is subject to the Coulomb potential

$$V_C(r) = -\frac{e^2}{4\pi\varepsilon_0 r} \tag{2.3.2}$$

due to the nucleus. Due to the spherical symmetry of the problem, it is convenient to write the wavefunction in spherical coordinates. From Eq. 2.1.2, it can be shown that

$$\psi_{nlm}(r, \theta, \phi) = R_{nl}(r)Y_{lm}(\theta, \phi).$$

The principal number $n = 1, 2, 3...$ gives the quantization of energy $E_n \propto -1/n^2$. The $R_{nl}(r)$ are *associated Laguerre functions* and determine the radial profile of the probability distribution for the electron. $Y_{lm}(\theta, \phi)$ are spherical harmonics, which can also be written in terms of the *associated Legendre functions*, $Y_{lm}(\theta, \phi) = P_l^m(\cos\theta)e^{im\phi}$. For $m = 0$, $Y_{l0}(\theta) = P_l(\cos\theta)$ are simply the Legendre polynomials. The azimuthal number $l = 0, 1, ..., n - 1$ labels the usual s, p, d, ... orbitals. The s orbitals are spherically symmetric, since $Y_{00}(\theta, \phi) = 1/\sqrt{4\pi}$. The higher the azimuthal number, the higher the probability to find the electron further away from the nucleus, whereas n gives the number of nodes of the wavefunction in the radial direction.

That the azimuthal quantum number l is quantized was demonstrated experimentally in what we now know as the *Zeeman effect*. In order to see why, we come back to the relation between the orbital angular momentum and the magnetic moment. We can write

$$m_{\mathrm{L}} = \mu_{\mathrm{B}}\sqrt{l(l+1)}$$
$$\mathbf{m}_{\mathrm{L}} \cdot \mathbf{z} = -\mu_{\mathrm{B}}m_l, \tag{2.3.3}$$

where $\mu_{\mathrm{B}} = \hbar\gamma_{\mathrm{L}}$ is the *Bohr magneton* and the expressions correspond to expectation values. If the atom is placed in an external magnetic field, there will be an extra contribution to the energy due to the Zeeman term, see Eq. 1.2.7. We can take the z-axis to coincide with the magnetic field, hence

$$E_Z = \mu_{\mathrm{B}}m_l B, \tag{2.3.4}$$

and we say that the degeneracy of the l level, originally $2l + 1$, is split for all $l > 0$. This splitting can be measured in the absorption spectrum of atoms with total spin angular momentum equal to zero. To add the effect of the spin degree of freedom, we have, however, first to understand how to combine angular momentum operators in quantum mechanics.

Check Points

- What are the quantum numbers for the hydrogen atom and what do they tell us?

2.4 Addition of Angular Momentum and Magnetic Moment

The orbital angular momentum is the generator of rotations in position space. In general, however, we can define an angular momentum simply by its algebra, determined by the commutation relation Eq. 2.2.2. The spin angular momentum generates rotations in spin space and satisfies

$$[\hat{S}_i, \hat{S}_j] = i\hbar\epsilon_{ijk}\hat{S}_k$$
$$\hat{S}^2|s\rangle = \hbar^2 s(s + 1)|s\rangle$$
$$\hat{S}_z|s\rangle = \hbar m_s|s\rangle.$$

In contrast to the orbital angular momentum, the spin quantum number s is not constrained to be an integer, and can take also half-integer values. Fermions (e.g., the electron) have half-integer values of spin and bosons integer values. For electrons, $s = 1/2$ and $m_s = \pm 1/2$.

The spin operator commutes with the orbital angular momentum

$$[\hat{\mathbf{S}}, \hat{\mathbf{L}}] = 0$$

since they act on different state spaces. One would be therefore tempted to choose $\left\{\hat{S}^2, \hat{L}^2, \hat{L}_z, \hat{S}_z\right\}$ as a set of commuting observables. Spin and orbital angular momentum, however, interact *via* the *spin–orbit interaction*, a relativistic correction to the Hamiltonian Eq. 2.1.4 which reads

$$\hat{H}_{\text{SO}} = \frac{\hbar^2}{2m_e^2 c^2}\frac{1}{r}\frac{\partial V}{\partial r}\hat{\mathbf{S}}\cdot\hat{\mathbf{L}}. \tag{2.4.1}$$

For an atomic system, V is the Coulomb potential. This correction is usually small, but it increases with the atomic number. Since $\hat{\mathbf{S}}$ and $\hat{\mathbf{L}}$ are coupled by Eq. 2.4.1, they are not conserved and they do not commute separately with the Hamiltonian. The total angular momentum $\hat{\mathbf{J}} = \hat{\mathbf{S}} + \hat{\mathbf{L}}$, however, is conserved. We choose therefore $\left\{\hat{S}^2, \hat{L}^2, \hat{J}^2, \hat{J}_z\right\}$ as a set of commuting observables, and s, l , j, and m_j are

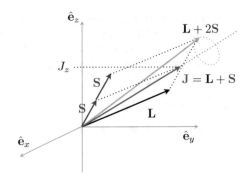

Fig. 2.2 Pictorial depiction
of the precession of
$\hat{\mathbf{L}} + 2\hat{\mathbf{S}} = \hat{\mathbf{m}}_{\text{TOT}}/\gamma_{\text{L}}$
around $\hat{\mathbf{J}}$

our quantum numbers. The change of basis is achieved via the *Clebsch–Gordan coefficients*

$$|sljm_j\rangle = \sum_{m_s m_l} |sljm_j\rangle\langle sm_s lm_l|sljm_j\rangle . \qquad (2.4.2)$$

We now have the task of relating the magnetic dipolar moment $\mathbf{m}_{\text{TOT}}$ to the total angular momentum operator $\hat{\mathbf{J}}$. From Eq. 1.4.4, we know that $\mathbf{m}_{\text{TOT}}$ is not collinear with $\hat{\mathbf{J}}$, but it is proportional to $\hat{\mathbf{L}} + 2\hat{\mathbf{S}}$. We note, however, that the projection $\hat{\mathbf{m}}_{\text{TOT}} \cdot \hat{\mathbf{J}}$ is well defined, since

$$[\hat{\mathbf{L}} \cdot \hat{\mathbf{J}}, \hat{\mathbf{J}}^2] = [\hat{\mathbf{L}} \cdot \hat{\mathbf{J}}, \hat{\mathbf{L}}^2] = [\hat{\mathbf{L}} \cdot \hat{\mathbf{J}}, \hat{\mathbf{S}}^2] = [\hat{\mathbf{L}} \cdot \hat{\mathbf{J}}, \hat{\mathbf{J}}_z] = 0$$
$$[\hat{\mathbf{S}} \cdot \hat{\mathbf{J}}, \hat{\mathbf{J}}^2] = [\hat{\mathbf{S}} \cdot \hat{\mathbf{J}}, \hat{\mathbf{L}}^2] = [\hat{\mathbf{S}} \cdot \hat{\mathbf{J}}, \hat{\mathbf{S}}^2] = [\hat{\mathbf{S}} \cdot \hat{\mathbf{J}}, \hat{\mathbf{J}}_z] = 0 . \qquad (2.4.3)$$

The magnetic moment therefore precesses around $\hat{\mathbf{J}}$, as depicted in Fig. 2.2.

The Wigner–Eckart theorem allows us to relate $\hat{\mathbf{m}}_{\text{TOT}}$ with $\hat{\mathbf{J}}$ in terms of expectation values by noting that [8]

$$\gamma_{\text{L}}\langle sljm_j|\hat{\mathbf{L}} + 2\hat{\mathbf{S}}|sljm'_j\rangle = \gamma_{\text{L}} g(slj)\langle sljm_j|\hat{\mathbf{J}}|sljm'_j\rangle .$$

Therefore, in the $(2j + 1)$ degenerate subspace with fixed quantum numbers (s, l, j) we can think of $\hat{\mathbf{m}}_{\text{TOT}}$ as being proportional to $\hat{\mathbf{J}}$. In the literature, this is sometimes denoted as "$\hat{\mathbf{m}}_{\text{TOT}} = \gamma_{\text{L}} g\hat{\mathbf{J}}$" in an abuse of notation. The value of $g(slj)$ we can find by projecting $\hat{\mathbf{m}}_{\text{TOT}} \cdot \hat{\mathbf{J}}$

$$\langle sljm_j|\left(\hat{\mathbf{L}} + 2\hat{\mathbf{S}}\right) \cdot \hat{\mathbf{J}}|sljm_j\rangle = \gamma_{\text{L}} g(slj)\langle sljm_j|\hat{\mathbf{J}}^2|sljm_j\rangle$$

and using Eqs. 2.4.3, we obtain the *Landé factor*

$$g(s, l, j) = \frac{3}{2} + \frac{s(s + 1) - l(l + 1)}{2j(j + 1)} . \qquad (2.4.4)$$

We see that this expression coincides with the classical one obtained in Eq. 1.5.11 in the limit of $s, l, j \gg 1$.

Going back to the Zeeman splitting, we see that in general the Zeeman correction to the atomic energy in the presence of a magnetic field is given by

$$E_Z = \mu_B g(s, l, j) m_j B, \tag{2.4.5}$$

and therefore the Zeeman splitting depends on all the atomic orbital numbers, and it is not simply $\mu_B B$. This is denominated the *anomalous Zeeman effect* since at the time of the experiments the spin was still not known, and it was not possible to explain the effect. Note, however, that the normal Zeeman effect can be observed only for atoms with zero total spin, and therefore the anomalous one is much more common.

1. **Exercise: Prove Eqs.** 2.4.3
2. **Exercise: Derive Eq.** 2.4.4.

Check Points

- How do you relate the magnetic moment of an electron to its total angular momentum? Why?
- What is the Landé factor?

2.5 Generalization to Many Electrons

Generalizing the previous concepts beyond the hydrogen atom/single electron problem is impossible to do in an exact manner, since the problem turns into a many-body problem: the many electrons interact not only with the nucleus, but among themselves. We can, however, make analytical progress by doing some reasonable approximations. The first one is called the *Hartree* approximation, in which we consider that each electron moves in an effective central potential $V_{eff}(r)$ generated by the nucleus plus all the other electrons. The other electrons are said to *screen* the potential of the nucleus, since their charge is opposite.

The second approximation concerns the spin–orbit coupling Eq. 2.4.1. For all except the heaviest atoms, the spin–orbit interaction is weak and can be treated within perturbation theory. In this case, we can first neglect this interaction and consider that $\hat{L}_i$ and $\hat{S}_i$ for each electron i are independent. Hence, we can calculate the total orbital and spin angular momentum simply by summing them separately: $\hat{L}_{TOT} = \sum_i \hat{L}_i$ and $\hat{S}_{TOT} = \sum_i \hat{S}_i$, and then proceed to calculate the total angular momentum $\hat{J}_{TOT} = \hat{L}_{TOT} + \hat{S}_{TOT}$ and the corresponding magnetic moment. $\hat{S}_{TOT}^2$ has eigenvalues $\hbar S(S + 1)$ with $S = \sum_i m_{s,i}$ and, respectively, $\hat{L}_{TOT}^2$ has eigenvalues $\hbar L(L + 1)$ with $L = \sum_i m_{l,i}$. The allowed values of $\hat{J}_{TOT}$ are given by the angular momentum summation rules: $J : |L - S|, |L - S| + 1, ...L + S$. This approximation is denominated the *Russell–Sanders coupling* and it gives rise to the well-known *Hund's rules*. To calculate $\hat{\mathbf{J}}_{TOT}$, we need first a prescription to obtain $\hat{\mathbf{S}}_{TOT}$ and $\hat{\mathbf{L}}_{TOT}$.

A closed shell means that we have occupied all $2(2l + 1)$ levels in it, where the factor of 2 comes for the spin $s = \pm 1$. Therefore, both the total orbital and spin angular momenta of the shell are zero, and hence also the total angular momentum $\hat{\mathbf{J}}$ in the shell. The total angular momentum of the atom, and therefore its magnetic properties, will be determined by the last, partially unoccupied shell. If all shells are closed (that is, full), then $\hat{\mathbf{J}}_{TOT} = 0$ and the atom is diamagnetic. The Hund rules tell us how to distribute our $n \leq 2(2l + 1)$ "leftover" electrons in the last shell. The total spin $\hat{\mathbf{S}}_{TOT}$ we obtain by applying the Pauli principle: since electrons are fermions, their wavefunction is antisymmetric and two electrons cannot have the same quantum numbers. Each orbital characterized by l can then be occupied by only two electrons: one with spin up, and one with spin down. The *first Hund rule* tells us to maximize S, since this will tend to put one electron within each orbital until half filling, $(2l + 1)$, and then continue with spin down. This minimizes Coulomb repulsion by putting electrons, in average, as far apart as possible. The *second Hund rule* tells us to maximize the orbital angular momentum, once the spin is maximized. This also minimizes Coulomb repulsion, by making the electrons orbit as far apart as possible. The *third Hund rule* sounds more mysterious: if the shell is more than half filled ($n \geq 2l + 1$) then $J = L + S$, and for a less than half-filled shell ($n \leq 2l + 1$), $J = |L - S|$. This is actually due to the spin–orbit interaction $\lambda \hat{\mathbf{L}}_{TOT} \cdot \hat{\mathbf{S}}_{TOT}$, where it can be shown that λ's sign changes between these two configurations. Therefore, to minimize spin–orbit coupling requires $\hat{\mathbf{S}}_{TOT}$ and $\hat{\mathbf{L}}_{TOT}$ parallel or antiparallel, depending on the filling.

To finish this section, we point out that for the heavier elements the Russell–Sanders coupling prescription is not valid anymore, due to the strong spin–orbit coupling. For these elements, a different prescription, denominated *jj coupling*, is used. There $\hat{\mathbf{J}}_i$ for each electron is first calculated, and then the total $\hat{\mathbf{J}}_{TOT}$.

As we mentioned, if all shells are closed, the total angular momentum is zero and the atom is diamagnetic. Diamagnetism can be understood by the Faraday law: as a magnetic field is turned on, it induces a change in the orbital motion of the electrons which opposes the change in magnetic flux. Diamagnetism is usually a weak effect and it is overshadowed by paramagnetism in atoms with partially unfilled shells. Once the atoms are ordered in a lattice and form a solid, it can happen that magnetic order develops as the temperature is lowered. This will depend on the electronic interactions, as we will see in the next chapter.

Check Points

• What is the Russell–Sanders coupling scheme?

Chapter 3
Magnetism in Solids

In the previous chapter, we showed how to calculate the magnetic moment of an atom. We saw that the problem is already quite involved even for a single atom if we go beyond a hydrogen-like one. When atoms come together to form a solid, to treat the magnetic problem atom per atom is not only impossible but also not correct, since we have to take into account the binding between the atoms that form the solid, and what matters is the collective behavior of the material. In a solid, the orbitals of the constituent atoms overlap to form bands instead of discrete energy levels. Depending on the character of the orbitals involved in the magnetic response of a material, we can divide the problem into two big subsets: metals and insulators. In metals, the orbitals are extended and have a good amount of overlap, so the electrons are *delocalized* and free to move around the solid. In insulators, the orbitals are narrow and we talk about *localized magnetic moments*. Of course, this is an oversimplified view: there are systems in which both localized and delocalized electrons participate in magnetism, or in which magnetism and electric conduction are due to different groups of orbitals. An example is that of *heavy fermion* systems, which can be modeled as a *Kondo Lattice*: a lattice of magnetic impurities embedded in a sea of free electrons. Itinerant magnetism (that one due to delocalized electrons) and mixed systems belong to what one calls *strongly correlated systems*, highly complex many-body problems in which electron–electron interactions have to be taken into account. In this course, we will concentrate mostly on magnetic insulators. Moreover, to understand *magnetic ordering*, we have to introduce electronic interactions and take into account the Pauli principle.

3.1 The Curie–Weiss Law

We consider first a system of localized, N identical noninteracting magnetic moments, and calculate their collective paramagnetic response.[1] We can obtain the magnetization from the Helmholtz free energy (see, e.g., Ref. [8])

$$F = -k_B T \ln Z \tag{3.1.1}$$

$$M = -\frac{N}{V}\frac{\partial F}{\partial B} \tag{3.1.2}$$

where k_B is the Boltzmann constant, T is the temperature, and Z is the canonical partition function for one magnetic moment

$$Z = \sum_n e^{-\frac{E_n}{k_B T}} . \tag{3.1.3}$$

If we consider a magnetic moment with total momentum J, we have $2J + 1$ possible J_z values and

$$Z = \sum_{J_z=-J}^{J} e^{-\frac{1}{k_B T}g\mu_B J_z B} . \tag{3.1.4}$$

The sum can be performed since it is a geometric one. Putting all together, one finds

$$M = \frac{N}{V}g\mu_B J \mathcal{B}_J(\frac{\mu_B g J B}{k_B T}), \tag{3.1.5}$$

where

$$\mathcal{B}_J(x) = \frac{2J+1}{2J}\coth(\frac{2J+1}{2J}x) - \frac{1}{2J}\coth(\frac{1}{2J}x)$$

is the Brillouin function. For $\mu_B B \gg k_B T$ this function goes to 1 and the magnetization saturates: all momenta are aligned with the B-field. For large temperatures instead, $\mu_B B \ll k_B T$, one obtains the inverse-temperature dependence of the magnetization known as the *Curie law*, characterized by the susceptibility

$$\chi_c = \frac{M}{H} = \frac{N}{V}\frac{\mu_0 (g\mu_B)^2}{3}\frac{J(J+1)}{k_B T} . \tag{3.1.6}$$

Experimentally, it is found that this result is good for describing insulating crystals containing rare-earth ions (e.g., Yb, Er), whereas for transition metal ions in an

---

[1] As we saw in the last chapter, all atoms present a diamagnetic behavior, but since it is very weak compared to the paramagnetic response, only atoms with closed shells (e.g., noble gases) present an overall diamagnetic response. An exception is of course superconducting materials, which can have a perfect diamagnetic response. This is, however, a collective, macroscopic response due to the superconducting currents which oppose the change in magnetic flux.

insulating solid agreement is found only if one takes $J = S$. This is due to an effect denominated *angular momentum quenching*, where effectively $L = 0$. This quenching is due to *crystal fields*: since now the atoms are located in a crystalline environment, rotational symmetry is broken and each atom is located in electric fields due to the other atoms in the crystal. In the case of rare-earth ions, the magnetic moments come from f-shells, which are located deep inside the atom and therefore better isolated form crystal fields. For transition metals instead, the magnetic moments come from the outermost d-shells, which are exposed to the symmetry breaking fields. These, however, are spatial dependent fields which do not affect directly the spin degree of freedom. As a consequence of the breaking of spatial rotational invariance, the orbital angular momentum is not conserved and precesses instead around the crystal fields, averaging to zero.

The Curie law is followed by all materials with net magnetic momentum for large enough temperature. For low temperatures, however, in certain materials, magnetic order develops and there is a deviation from the Curie law in the susceptibility. Weiss postulated the existence of *molecular fields*, which are proportional to the magnetization in the material and are responsible for the magnetic ordering. The total magnetic field acting on a magnetic moment within this picture is, therefore, $\mathbf{H}_{tot} = \mathbf{H}_{ext} + \lambda \mathbf{M}$ and from the Curie susceptibility

$$\chi_c = \frac{M}{H_{tot}} = \frac{M}{H_{ext} + \lambda M} = \frac{C}{T} \tag{3.1.7}$$

with

$$C = \frac{N}{V} \frac{\mu_0 (g\mu_B)^2}{3} \frac{J(J+1)}{k_B} \tag{3.1.8}$$

we obtain

$$M = \frac{C H_{ext}}{T - \lambda C} \tag{3.1.9}$$

and therefore the susceptibility in the external field follows the *Curie–Weiss law*

$$\chi_w = \frac{C}{T - \lambda C}. \tag{3.1.10}$$

We see that for large temperatures this susceptibility follows the Curie inverse law, but as the temperature approaches a critical temperature $T_C = \lambda C$, χ_w diverges indicating a phase transition to the magnetically ordered phase.

A magnetically ordered phase implies $M \neq 0$ for an external field $B_{ext} = 0$. From Eq. 3.1.5 we see that, if we do not consider the molecular fields postulated by Weiss, $M(0) = 0$. Let us now consider $B_{ext} = 0$ but the existence of a molecular field $B_W = \mu_0 \lambda M$. Inserting this field in Eq. 3.1.5, we obtain an implicit equation for M:

$$M = M_0 \mathcal{B}_J \left(\frac{\mu_B \mu_0 g J \lambda M}{k_B T} \right), \tag{3.1.11}$$

Fig. 3.1 Possibility of spontaneous magnetic order according to the condition Eq. 3.1.13. For temperatures lower than the critical temperature T_C, there is a nontrivial solution of Eq. 3.1.11, indicating magnetic order with a spontaneous magnetization M_{sp}. The inset depicts schematically the behavior of the spontaneous magnetization M_{sp} with temperature

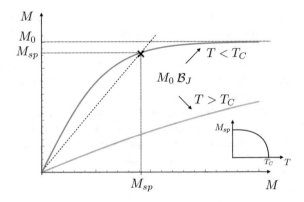

where we have defined the saturation magnetization

$$M_0 = \frac{N}{V} g \mu_B J \tag{3.1.12}$$

since $\mathcal{B}_J(x \to \infty) = 1$. Equation 3.1.11 has still a solution $M = 0$. However, if

$$\left[\frac{\mathrm{d}(M_0 \mathcal{B}_J)}{\mathrm{d}M} \right]_{M=0} \geq 1, \tag{3.1.13}$$

we see that a second solution to the transcendental equation is possible. From

$$\mathcal{B}_J(x \to 0) \approx \frac{J+1}{3J} x,$$

we obtain

$$M_{sp} = \lambda C \frac{M_{sp}}{T_C}$$

and therefore $T_C = \lambda C$, in agreement with Eq. 3.1.10. The subscript sp indicates this is a *spontaneous magnetization*, not induced by an external magnetic field. For $T > T_C$, Eq. 3.1.13 is not fulfilled and $M = 0$ is the only solution. This can be seen graphically in Fig. 3.1.

At the time, the origin of these postulated molecular fields was not known. Naively, one could expect the dipole–dipole interaction between the magnetic moments to be the origin of the magnetic ordering. This energy scale is, however, too small to explain magnetism at room temperature. The potential energy of one magnetic dipole $\mathbf{m}_2$ in the magnetic field created by another dipole $\mathbf{m}_1$, $V_{pot} = -\mathbf{m}_2 \cdot \mathbf{B}_1(\mathbf{r})$ is

$$V_{pot} = -\mathbf{m}_2 \cdot \frac{\mu_0}{4\pi\varepsilon_0} \frac{3(\mathbf{m}_1 \cdot \mathbf{r})\mathbf{r} - r^2 \mathbf{m}_1}{r^5}. \tag{3.1.14}$$

A quick estimate corresponds to taking the distance between the dipoles as the interatomic distance, and the dipolar moments simply as Bohr magnetons. Equating this energy to $k_B T_C$ results in a critical temperature for magnetic ordering of $T_C \approx 1\,K$. Therefore, dipolar interactions could be responsible magnetic ordering only below this temperature, which is very low. We know, however, that magnetic ordering at room temperature is possible. Instead, a rough estimate of the repulsive Coulomb energy between two electrons gives

$$\frac{U_c}{k_B} = \frac{e^2}{k_B 4\pi\varepsilon_0 a^2} \approx 10^5\,K$$

which is very large! This could provide us with the necessary energy scale. In the following section, we will see that magnetic ordering is due to a combination of the electrostatic energy and a very quantum effect: the Pauli principle.

1. **Exercise: Prove Eq.** 3.1.5.
2. **Exercise: estimate T_C from the dipole–dipole interaction.**

Check Points

- What is the Curie law and when is it valid?
- What is the Curie–Weiss law?
- Why the dipolar–dipolar interaction cannot in general explain magnetic ordering?

3.2 Exchange Interaction

Let us consider first the case of two electrons subject to a Hamiltonian

$$H = \frac{p_1^2}{2m_e} + \frac{p_2^2}{2m_e} + V(\mathbf{r}_1, \mathbf{r}_2). \tag{3.2.1}$$

This Hamiltonian is independent of spin; however, the Pauli exclusion principle imposes a spatial symmetry on the wavefunction $\psi(\mathbf{r}_1, \mathbf{r}_2)$ solution of $H\psi(\mathbf{r}_1, \mathbf{r}_2) = E\psi(\mathbf{r}_1, \mathbf{r}_2)$. Since two electrons with the same quantum numbers are not allowed at the same place, the total wavefunction $\psi(\mathbf{r}_1, \mathbf{r}_2; s_1, s_2)$ must be antisymmetric with respect to exchange of both spin and space:

$$\psi(\mathbf{r}_1, \mathbf{r}_2; s_1, s_2) = -\psi(\mathbf{r}_2, \mathbf{r}_1; s_2, s_1).$$

If there is no spin–orbit coupling in our Hamiltonian, we can separate the wavefunction

$$\psi(\mathbf{r}_1, \mathbf{r}_2; s_1, s_2) = \psi(\mathbf{r}_1, \mathbf{r}_2)\chi(s_1, s_2)$$

and therefore we see that the antisymmetry implies a correlation between the spin and orbital parts of the wavefunction. This is a constraint of the problem, and solutions that do not fulfill this constraint are infinite in energy. For two electrons, we have the total $S = 0, 1$ and correspondingly $S_z = 0, -1, 0, 1$. We can write the corresponding states in explicit symmetric and antisymmetric linear combinations, the antisymmetric singlet state

$$|0; 0\rangle \frac{1}{\sqrt{2}} (|\uparrow\downarrow\rangle - |\downarrow\uparrow\rangle) \tag{3.2.2}$$

and the symmetric triplet state with $S = 1$, $S_z = -1, 0, 1$:

$$|1; 1\rangle = |\uparrow\uparrow\rangle \tag{3.2.3}$$

$$|1; 0\rangle = \frac{1}{\sqrt{2}} (|\uparrow\downarrow\rangle + |\downarrow\uparrow\rangle)$$

$$|1; -1\rangle = |\downarrow\downarrow\rangle$$

where the states are labeled as $|S; S_z\rangle$. Since the Hamiltonian does not depend on spin, we can work simply with the spatial component of the wavefunction, but imposing the right symmetry. The spatial part of the wavefunction corresponding to the singlet configuration, $\psi_{singlet}(\mathbf{r}_1, \mathbf{r}_2)$ will therefore be symmetric in space coordinates, whereas $\psi_{triplet}(\mathbf{r}_1, \mathbf{r}_2)$ has to be antisymmetric

$$\psi_{singlet}(\mathbf{r}_1, \mathbf{r}_2) = \psi_{singlet}(\mathbf{r}_2, \mathbf{r}_1)$$

$$\psi_{triplet}(\mathbf{r}_2, \mathbf{r}_1) = -\psi_{triplet}(\mathbf{r}_2, \mathbf{r}_1) , \tag{3.2.4}$$

and we can write the full state as

$$|\Psi_{singlet}\rangle = |\psi_{singlet}\rangle |0; 0\rangle$$

$$|\Psi_{triplet}(S_z)\rangle = |\psi_{triplet}\rangle |1; S_z.\rangle \tag{3.2.5}$$

The Schrödinger equation therefore reads

$$H|\psi_{singlet}\rangle = E_s|\psi_{singlet}\rangle$$

$$H|\psi_{triplet}(S_z)\rangle = E_t|\psi_{triplet}\rangle$$

and if $E_s \neq E_t$, the ground state is spin-dependent even though H is not. If that is the case, since spatial and spin sectors are correlated, we search for a Hamiltonian operating in spin space $\tilde{H}$, coupling $\hat{s}_1$ and $\hat{s}_2$ and such that it is equivalent to H:

$$\tilde{H}|0; 0\rangle = E_s|0; 0\rangle \tag{3.2.6}$$

$$\tilde{H}|1; S_z\rangle = E_t|1; S_z\rangle .$$

The Hamiltonian that does the trick is [3]

$$\tilde{H} = \frac{1}{4} (E_s + 3E_t) - \frac{1}{\hbar^2} (E_s - E_t) (\hat{\mathbf{s}}_1 \cdot \hat{\mathbf{s}}_2) \tag{3.2.7}$$

since

$$\frac{\hat{\mathbf{s}}_1 \cdot \hat{\mathbf{s}}_2}{\hbar^2} = \frac{1}{2} S(S+1) - \frac{3}{4} = \begin{cases} -\frac{3}{4} & S = 0 \\ \frac{1}{4} & S = 1 \end{cases}. \tag{3.2.8}$$

We have therefore constructed a Hamiltonian acting on spin space which is in principle equivalent to the interacting Hamiltonian of Eq. 3.2.1, which gives an effective interaction between the spins. This is called the *molecular Heisenberg model* and can be written as

$$\tilde{H} = J_0 - J_{12}\hat{\mathbf{s}}_1 \cdot \hat{\mathbf{s}}_2$$

with

$$J_{12} = \frac{1}{\hbar^2} (E_s - E_t) . \tag{3.2.9}$$

If $J_{12} > 0$, this interaction favors a ferromagnetic alignment of the spins, consistent with the fact that the singlet energy is higher than the triplet one. We now show an example in which $E_s \neq E_t$ and calculate explicitly J_{12}. We will see that part of J_{12} has no classical analog and comes from interchanging particles 1 and 2, since in quantum mechanics they are indistinguishable.

1. **Exercise: Prove Eq.** 3.2.8 **and show that** $\tilde{H}$ **given in Eq.** 3.2.7 **satisfies** 3.2.6.

Check Points

- Why can you write a Hamiltonian in spin space which is equivalent to the Hamiltonian defined in position space?
- How do you impose the equivalence for the two-electron system?

3.3 Hydrogen Molecule

We consider now two hydrogen atoms that are brought close together to form a hydrogen molecule. We consider the nuclei, a and b, as fixed at positions $\mathbf{R}_a$ and $\mathbf{R}_b$, whereas the electrons, at positions $\mathbf{r}_1$ and $\mathbf{r}_2$, are subject to the nuclei Coulomb potential plus the repulsive Coulomb interaction between them . This corresponds to taking the nuclei mass $m_a, m_b \to \infty$ and it is an example of the denominated *Born Oppenheimer approximation*. The Hamiltonian for each, separate hydrogen atom a and b are given by

$$H_a = -\frac{\hbar \nabla_1^2}{2m_e} - \frac{e^2}{4\pi\varepsilon_0} \frac{1}{|\mathbf{R}_a - \mathbf{r}_1|}$$

$$H_b = -\frac{\hbar \nabla_2^2}{2m_e} - \frac{e^2}{4\pi\varepsilon_0} \frac{1}{|\mathbf{R}_b - \mathbf{r}_2|}$$

and we know the respective eigenfunctions, $\phi_{a,b}$ with eigenenergies $E_{a,b}$. If the distance between the two atoms $|\mathbf{R}_a - \mathbf{R}_b| \to \infty$, these are the exact solutions. If, however, the atoms are brought close together to form a molecule, there will be an interaction term

$$H_I = \frac{e^2}{4\pi\varepsilon_0} \frac{1}{|\mathbf{R}_a - \mathbf{R}_b|} - \frac{e^2}{4\pi\varepsilon_0} \frac{1}{|\mathbf{R}_b - \mathbf{r}_1|} - \frac{e^2}{4\pi\varepsilon_0} \frac{1}{|\mathbf{R}_a - \mathbf{r}_2|} + \frac{e^2}{4\pi\varepsilon_0} \frac{1}{|\mathbf{r}_2 - \mathbf{r}_1|}$$

and the total Hamiltonian is given by

$$H_{tot} = H_a + H_b + H_I \,.$$

Even though this is a quite simple system, this problem cannot be solved exactly and we have to resort to approximations. We treat H_I as a perturbation and use the atomic wavefunctions $\phi_{a,b}$ as a basis for a variational solution of the full wavefunction. This implies we are assuming the electrons are quite localized at their respective atom. This approach is denominated the *Heitler–London* method.

If we consider the unperturbed Hamiltonian

$$H_0 = H_a + H_b$$

the eigenfunctions will be simply linear combinations of the product of the original orbitals, $\phi_a \phi_b$, with eigenenergy $E_a + E_b$ since the two systems do not interact with each other. To preserve the indistinguishability of the particles, we cannot simply write a solution as $\phi_a(\mathbf{r}_1)\phi_b(\mathbf{r}_2)$, since the probability density $|\phi_a(\mathbf{r}_1)|^2|\phi_b(\mathbf{r}_2)|^2$ is not invariant under exchanging $\mathbf{r}_1 \longleftrightarrow \mathbf{r}_2$. The symmetric and antisymmetric combinations fulfill, however, the indistinguishability condition

$$\psi_s(\mathbf{r}_1, \mathbf{r}_2) = \frac{1}{\sqrt{2}} \left(\phi_a(\mathbf{r}_1)\phi_b(\mathbf{r}_2) + \phi_b(\mathbf{r}_1)\phi_a(\mathbf{r}_2) \right)$$

$$\psi_t(\mathbf{r}_1, \mathbf{r}_2) = \frac{1}{\sqrt{2}} \left(\phi_a(\mathbf{r}_1)\phi_b(\mathbf{r}_2) - \phi_b(\mathbf{r}_1)\phi_a(\mathbf{r}_2) \right), \qquad (3.3.1)$$

where with the notation ψ_s, ψ_t we have anticipated that the symmetric (antisymmetric) solution in space corresponds to the singlet (triplet) solution in spin space, with a full wavefunction as given in Eq. 3.2.5. Note that both ψ_s, ψ_t are eigenfunctions of H_0 with eigenvalue $E_a + E_b$, and therefore without the interaction H_I, $E_t = E_s = E_a + E_b$ and the solutions are fourfold degenerate.

We now calculate E_t and E_s perturbatively in the presence of the interaction H_I and using ψ_s, ψ_t as variational wavefunctions

$$E_{s/t} = \frac{\langle \psi_{s/t} | H_{tot} | \psi_{s/t} \rangle}{\langle \psi_{s/t} | \psi_{s/t} \rangle} . \tag{3.3.2}$$

We are interested in the ground state solutions, so that $\phi_{a,b}$ are solutions of each hydrogen atom with $E_{a,b} = E_0$. The variational principle tells us that the energies calculated by Eq. 3.3.2 are always greater or equal that the true ground state.[2]

Let us first analyze the simple overlap

$$\langle \psi_{s/t} | \psi_{s/t} \rangle = \int d^3 r_1 d^3 r_2 \psi_{s/t}^*(\mathbf{r}_1, \mathbf{r}_2) \psi_{s/t}(\mathbf{r}_1, \mathbf{r}_2) . \tag{3.3.3}$$

If the two atoms are infinitely apart $|\mathbf{R}_a - \mathbf{R}_b| \to \infty$, then the orbitals corresponding to different atoms have zero overlap, $\langle \phi_a | \phi_b \rangle = 0$, and

$$\langle \psi_{s/t} | \psi_{s/t} \rangle_0 = \int d^3 r_1 d^3 r_2 |\phi_a(\mathbf{r}_1)|^2 |\phi_b(\mathbf{r}_2)|^2 = 1 . \tag{3.3.4}$$

When the atoms are brought close together, their orbitals will overlap: $\langle \phi_a | \phi_b \rangle \neq 0$. We do not calculate this overlap explicitly, but simply note that it will be finite and denote it by O

$$O^2 = \int d^3 r_1 d^3 r_2 \phi_a^*(\mathbf{r}_1) \phi_b(\mathbf{r}_1) \phi_a(\mathbf{r}_2) \phi_b^*(\mathbf{r}_2) \tag{3.3.5}$$

and hence

$$\langle \psi_{s/t} | \psi_{s/t} \rangle = 1 \pm O^2 . \tag{3.3.6}$$

We now turn to the numerator in Eq. 3.3.2. We already know that the noninteracting contribution is

$$\frac{\langle \psi_{s/t} | H_0 | \psi_{s/t} \rangle}{\langle \psi_{s/t} | \psi_{s/t} \rangle} = 2E_0 . \tag{3.3.7}$$

The correction to this noninteracting energy is given by the term containing $\langle \psi_{s/t} | H_I | \psi_{s/t} \rangle$, which we see contains two kinds of terms. One of them is simply the Coulomb electrostatic interaction between the two atoms, assuming that they are close enough to interact but electron 1(2) still "belongs" to atom $a(b)$

$$K = \int d^3 r_1 d^3 r_2 \phi_a^*(\mathbf{r}_1) \phi_b(\mathbf{r}_2) H_I \phi_a(\mathbf{r}_1) \phi_b^*(\mathbf{r}_2) . \tag{3.3.8}$$

---

[2]Note that in general the variational procedure would be to write $\psi_{s/t}(\mathbf{r}_1, \mathbf{r}_2) = c_1 \phi_a(\mathbf{r}_1) \phi_b(\mathbf{r}_2) \pm c_2 \phi_b(\mathbf{r}_1) \phi_a(\mathbf{r}_2)$ and find the coefficients $c_{1,2}$ by minimizing Eq. 3.3.2. One then finds $c_1 = c_2 = 1/\sqrt{2}$. In Eq. 3.3.1 we used our knowledge of the symmetry of the problem plus the normalization of the wavefunctions $\phi_{a,b}$ to write the result immediately.

The remaining term has no classical analog, and it measures the Coulomb energy cost upon exchanging the two electrons

$$X = \int d^3r_1 d^3r_2 \phi_a^*(\mathbf{r}_1)\phi_b(\mathbf{r}_1)H_I\phi_a(\mathbf{r}_2)\phi_b^*(\mathbf{r}_2) \,. \tag{3.3.9}$$

and it is called the *exchange integral*, or *exchange interaction*. Putting all together, we obtain

$$E_{s/t} = 2E_0 + \frac{K \pm X}{1 \pm O^2}, \tag{3.3.10}$$

where the $+$ $(-)$ corresponds to the singlet (triplet) solution ψ_s (ψ_t). In general, $O \ll 1$ and we can replace the denominator by 1. We have therefore shown that $E_s - E_t \neq 0$ and hence for this problem

$$J_{12} \approx \frac{2}{\hbar^2} X \,, \tag{3.3.11}$$

which justifies the name *exchange parameter* for J_{12}.

Check Points

- What is the Heitler–London model?
- What is the exchange interaction and why does it not have a classical analog?

3.4   Heisenberg, Ising, and XY Models

In our variational solution for the hydrogen molecule from the last section, double occupation is forbidden, so two electrons cannot be in the same atom at the same time. This indicates our treatment is valid for insulators, where electrons are quite localized, but in turn leads necessarily to small values of the exchange parameter, since J_{12} relies on the overlap of the single-atom orbitals. The above example therefore must be taken as a toy model which reveals the character of the ferromagnetic interaction. In general, the exchange constant is generated by more complex interactions, *e.g.*, *superexchange* where the ferromagnetic exchange interaction between two spins is mediated by an exchange interaction with an atom in between those with a net angular momentum.

We further postulate that our model can be generalized to N multielectron atoms

$$H = -\frac{1}{2} \sum_{ij} J_{ij} \hat{\mathbf{S}}_i \cdot \hat{\mathbf{S}}_j, \tag{3.4.1}$$

where the exchange coefficient J_{ij} is taken as a parameter of the model that has to be calculated for each particular material. The Hamiltonian in Eq. 3.4.1 is the

Heisenberg Hamiltonian. The factor of $1/2$ accounts for the double-counting in the sum. We write $\hat{\mathbf{S}}$ by convention, and we refer to the magnetic moment as "spins" in an abuse of language: in reality, unless the orbital angular momentum is quenched, the total angular momentum of the ions is meant. The spin operators follow the angular momentum algebra when located at the same site, and commute with operators at different sites:

$$[\hat{S}_i^\alpha, \hat{S}_j^{\ \beta}] = i\hbar\delta_{ij}\epsilon_{\alpha\beta\gamma}\hat{S}_i^\gamma, \tag{3.4.2}$$

where α, β, and γ indicate the spatial components of the angular momentum x, y, and z. These commutation relations make the quantum Heisenberg model, despite its simple appearance, quite a rich model, and exactly solvable only for a few simple cases. The input of the model is the lattice connectivity and dimensionality, and the exchange parameter J_{ij}.

Besides insulators, the Heisenberg Hamiltonian is a valid model for localized magnetic moments embedded in a metal. In that case, the exchange interaction is mediated by the conduction electrons, which gives rise to the *RKKY interaction* (Ruderman–Kittel–Kasuya–Yosida). The calculated exchange function J_{ij} is an oscillating function of position, alternating between positive and negative values, and is longer ranged than in the insulating case.

If there are no local magnetic moments but the system still presents magnetic order, the conduction electrons are also responsible for the magnetic order. In this case, the magnetism is denominated *itinerant* and it is described by a different Hamiltonian: the *Hubbard Hamiltonian*. This model takes into account the kinetic energy of the electrons, who can "jump" from lattice site to lattice site, and penalizes double occupation with a local Coulomb repulsion term. The Hubbard model takes an effective Heisenberg form in the particular case of a half-filled band and strong Coulomb interaction.

The Heisenberg Hamiltonian is the "father" Hamiltonian of other well-known models in magnetism. In a crystal, crystal fields can give rise to anisotropies in the exchange parameter J_{ij}. If the anisotropy is along only one direction, one can write

$$H = -\sum_{ij} \tilde{J}_{ij}\left(\hat{S}_i^x\hat{S}_j^x + \hat{S}_i^y\hat{S}_j^y + \Delta\,\hat{S}_i^z\hat{S}_j^z\right). \tag{3.4.3}$$

If $\Delta > 1$, magnetic ordering occurs along the z-axis, which is denominated the *easy axis*. For $\Delta \gg 1$, the Hamiltonian turns effectively into the *Ising Model*

$$H_{\text{Ising}} = -\sum_{ij} J_{ij}\,S_i^z S_j^z. \tag{3.4.4}$$

Note that in this particular case, all operators in the Hamiltonian commute, and therefore the model is in this sense classical. If $\Delta < 1$, then we have an *easy plane* ordering. For $\Delta \ll 1$, the system is effectively two-dimensional and isotropic, which is termed the *XY* model

$$H_{\mathrm{XY}} = -\sum_{ij} J_{ij} \left(\hat{S}_i^x \hat{S}_j^x + \hat{S}_i^y \hat{S}_j^y \right) . \qquad (3.4.5)$$

If the system is placed in an external magnetic field, a Zeeman term is added to the Hamiltonian

$$\begin{aligned}
H &= -\frac{1}{2} \sum_{ij} J_{ij} \hat{\mathbf{S}}_i \cdot \hat{\mathbf{S}}_j - g\mu_{\mathrm{B}} \mathbf{B} \cdot \sum_i \hat{\mathbf{S}}_i \\
&= -\frac{1}{2} \sum_{ij} J_{ij} \hat{\mathbf{S}}_i \cdot \hat{\mathbf{S}}_j - g\mu_{\mathrm{B}} B \hat{S}_i^z .
\end{aligned} \qquad (3.4.6)$$

In Eq. (3.4.6), the spin operators are dimensionless and $\hbar$ has been absorbed in μ_{B} (correspondingly, in the commutation relation Eq. (3.4.2) $\hbar$ should be set to 1 if this convention is used). Note that the Zeeman energy is $-\mathbf{m} \cdot \mathbf{B}$ (see Eq. (1.2.7)) and tends to align the magnetic moment with the magnetic field, and anti-align the angular momentum. Sometimes, by convention the extra minus sign is not used, since from now onwards one always works with the angular momentum operators, and one takes $g\mu_{\mathrm{B}} > 0$. This corresponds simply to transform $\mathbf{B} \to -\mathbf{B}$ if we want to translate into the magnetization or magnetic moments, and does not affect the results.

Check Points

- Write the Heisenberg Hamiltonian in a magnetic field.

3.5 Mean Field Theory

Once we have a Hamiltonian that models our system, we want in principle to (i) find the ground state, that is, the lowest energy eigenstate of the system, which is the only state populated at zero temperature, (ii) find the excitations on top of this ground state, which will determine the behavior of the system at $T \neq 0$, and (iii) study phase transitions, either at $T = 0$ (denominated a *quantum phase transition*) whereby changing some other external parameter like the magnetic field, the ground state of the system changes abruptly, or at finite temperature, where an order parameter of the system (given by a quantum-statistical average of some relevant quantity to describe the system) goes to zero as a function of temperature or other external parameters. An example of the latter is the magnetization $M(B, T)$. We go back now to the issue of magnetic ordering armed with the Heisenberg Hamiltonian. As we pointed out above, this is a very rich model, and there are very few general statements that can be made about the three points mentioned above. We turn therefore first to a well-known approximation denominated *mean field theory*. This approximation is in general good only for long-range interactions and high dimensions; it is, however, widely used to get a first idea of, for example, what kind of phases and phase transitions our model can present. For the Heisenberg model, we will see that mean field theory will give as a microscopic justification of the Weiss molecular fields we introduced at the beginning of this chapter.

We start by assuming that $\langle \hat{\mathbf{S}}_i \rangle$ is finite. For example, for a ferromagnetic ground state, $\langle \hat{\mathbf{S}}_i \rangle$ is uniform and such that

$$\mathbf{M} = \frac{N}{V} g\mu_B \langle \hat{\mathbf{S}}_i \rangle, \tag{3.5.1}$$

where N is the number of lattice sites and V is the total volume. We write now $\hat{\mathbf{S}}_i$ in the suggestive form

$$\hat{\mathbf{S}}_i = \langle \hat{\mathbf{S}}_i \rangle + \left(\hat{\mathbf{S}}_i - \langle \hat{\mathbf{S}}_i \rangle \right), \tag{3.5.2}$$

which corresponds to splitting the operator into its quantum-statistical average value and the fluctuations with respect to this average. In mean field theory, these fluctuations are assumed to be small. The *mean field Hamiltonian* is obtained from the Heisenberg Hamiltonian by keeping only terms up to first order in the fluctuation,

$$H_{\mathrm{MF}} = \frac{1}{2} \sum_{ij} J_{ij} \langle \hat{\mathbf{S}}_i \rangle \cdot \langle \hat{\mathbf{S}}_j \rangle - \sum_{ij} J_{ij} \langle \hat{\mathbf{S}}_i \rangle \cdot \hat{\mathbf{S}}_j - g\mu_B \mathbf{B} \cdot \sum_i \hat{\mathbf{S}}_i, \tag{3.5.3}$$

where we already included an external magnetic field $\mathbf{B}$. The first term in Eq. (3.5.3) is simply a constant shift in the energy. The second term corresponds to a spin a site i in the presence of a magnetic field generated by all other spins. We can therefore define an effective magnetic field

$$\mathbf{B}_{\mathrm{eff}} = \mathbf{B} + \frac{1}{g\mu_B} \sum_i J_{ij} \langle \hat{\mathbf{S}}_i \rangle \tag{3.5.4}$$

and our problem is reduced from an interacting problem (where spins interact with each other via the exchange interaction), to that of noninteracting spins in the presence of a magnetic field. If we go back to Eq. (3.1.7), we see that now we can give a microscopic explanation to the Weiss molecular fields, which were assumed to be proportional to to the magnetization. In particular, we find that we can write

$$\mathbf{B}_{\mathrm{eff}} = \mathbf{B} + \frac{\lambda}{\mu_0} \mathbf{M} \tag{3.5.5}$$

with

$$\lambda = \frac{1}{\mu_0} \frac{V}{N} \frac{1}{(g\mu_B)^2} J_0, \tag{3.5.6}$$

where we have used translational symmetry and defined $J_0 = \sum_i J_{ij}$ independent of j. This system therefore presents a transition to an ordered state as a function of temperature and magnetic field as discussed for the Curie–Weiss law, but we now have a microscopic explanation for the phenomenological model.

Check Points

- What is the meaning of the mean field theory?
- How is it related to the Curie–Weiss law?

3.6 Ground State of the Ferromagnetic Heisenberg Hamiltonian

For the particular case in which $J_{ij} \geq 0 \,\forall\, i, \ j$ it is possible to find the ground state of the Heisenberg Hamiltonian without further specifications. This is called the *ferromagnetic Heisenberg model* since the ground state is ferromagnetically ordered, as we show in the following.

We consider, hence, the Hamiltonian

$$\hat{H} = -\frac{1}{2} \sum_{ij} J_{ij} \hat{\mathbf{S}}_i \cdot \hat{\mathbf{S}}_j \ , \ \text{with } J_{ij} = J_{ji} \geq 0. \tag{3.6.1}$$

If the spins were classical vectors, the state of lowest energy would be that one with all N spins are aligned. Hence, a natural candidate for the ground state of $\hat{H}$ is

$$|0\rangle = |S, S\rangle_1 |S, S\rangle_2 ... |S, S\rangle_N, \tag{3.6.2}$$

where $\hat{S}_i^2 |S, S\rangle_i = S(S+1)|S, S\rangle_i$ and $\hat{S}_i^z |S, S\rangle_i = S|S, S\rangle_i$, that is, all spins take their maximum projection of $\hat{S}^z$ (we consider all spins identical). The individual spin operators, however, do not commute with the Hamiltonian 3.6.1,

$$\left[\hat{\mathbf{S}}_i, \hat{H} \right] \neq 0$$

and therefore a product states of the form

$$|\psi\rangle = |S, m_1\rangle |S, m_2\rangle ... |S, m_N\rangle \tag{3.6.3}$$

(with $\hat{S}_i^z |S, m_i\rangle = m_i |S, m_i\rangle$) span a basis of the Hilbert space, but are not necessarily eigenstates of $\hat{H}$. The total spin operator $\hat{\mathbf{S}}_{\text{TOT}}$, however, does commute with $\hat{H}$

$$\left[\hat{\mathbf{S}}_{\text{TOT}}, \hat{H} \right] = 0$$

and we can construct an eigenbasis for $\hat{H}$, $\hat{S}^2_{TOT}$, and $\hat{S}^z_{TOT}$. The state 3.6.2 is a state with maximum $\hat{S}_{TOT}$. We will now prove that (i) 3.6.2 is an eigenstate of 3.6.1, and (ii) that there is no eigenstate with higher energy [8].

To follow with the proof, it is convenient to recast $\hat{H}$ in term of *ladder operators*

$$\hat{S}^{\pm}_i = \hat{S}^x_i \pm i\hat{S}^y_i \tag{3.6.4}$$

$$\hat{S}^{\pm}_i |S_i, m_i\rangle = \sqrt{(S_i \mp m_i)(S_i + 1 \pm m_i)}|S_i, m_i \pm 1\rangle,$$

from which it is clear why they are call ladder operators: $\hat{S}^+$ ($\hat{S}^-$) increases (decreases) the projection of $\hat{S}^z$ by one unit. In terms of 3.6.4, the Hamiltonian 3.6.1 reads

$$\hat{H} = -\frac{1}{2}\sum_{ij} J_{ij} \left[\frac{1}{2}\left(\hat{S}^+_i \hat{S}^-_j + \hat{S}^-_i \hat{S}^+_j \right) + \hat{S}^z_i \hat{S}^z_j \right]. \tag{3.6.5}$$

With this expression, it is now straightforward to show that $|0\rangle$ is an eigenstate of $\hat{H}$

$$\hat{H}|0\rangle = -\frac{1}{2}\sum_{ij} J_{ij}\hat{S}^z_i \hat{S}^z_j|0\rangle = -\frac{S^2}{2}\sum_{ij} J_{ij}|0\rangle$$

since $\hat{S}^+_i|0\rangle = 0 \,\forall\, i$ (remember that spin operators at different sites commute, and $J_{ii} = 0$). Therefore,

$$\hat{H}|0\rangle = E_0|0\rangle$$

with

$$E_0 = -\frac{S^2}{2}\sum_{ij} J_{ij}.$$

We now need to prove that E_0 is the minimum possible energy. For that, we consider the expectation value of $\hat{H}$ with an arbitrary product state as in Eq. 3.6.3

$$E'_0 = \langle\psi|\hat{H}|\psi\rangle = -\frac{1}{2}\sum_{ij} J_{ij}\langle\psi|\hat{S}^z_i \hat{S}^z_j|\psi\rangle = -\frac{1}{2}\sum_{ij} J_{ij}m_i m_j$$

where we have used that

$$\hat{S}^+_i \hat{S}^-_j|\psi\rangle \propto |S, m_1\rangle...|S, m_i + 1\rangle...|S, m_j - 1\rangle...|S, m_N\rangle$$

and therefore

$$\langle\psi|\hat{S}^+_i \hat{S}^-_j|\psi\rangle = 0.$$

We further note that if $J_{ij} \geq 0$ then

$$\sum_{ij} J_{ij} m_i m_j \leq \sum_{ij} J_{ij} SS$$

and hence

$$E_0' \geq E_0,$$

which proves that the fully polarized state $|0\rangle$ given in Eq. 3.6.2 is indeed the ground state. This ground state is, however, not unique, but it is $(2S_{TOT} + 1)$ degenerate in spin space, with $S_{TOT} = NS$. This can be easily visualized in the two-spin case, where for the triplet state, $S_{TOT} = 1$ and the state has three possible projections of $\hat{S}_z$, see Eq. 3.2.3. Note that each of these states is, moreover, infinitely degenerate in position space, since we are able to choose the quantization axis freely. This is an example of what is called *spontaneous symmetry breaking*, which occurs when the ground state has a lower symmetry than the Hamiltonian [9]. The ground state of $\hat{H}$ is a specific realization of the $(2S_{TOT} + 1)$ possible ground states, and therefore has "picked" a preferred direction in spin space, which is not determined by the symmetry of $\hat{H}$. We mention in passing that our results are valid in all its generality strictly for $T = 0$. For $T > 0$, the *Mermin–Wagner theorem* tells us that, in one and two dimensions and for short-range interactions, continuous symmetries cannot be spontaneously broken.

If we add a magnetic field to $\hat{H}$, our total Hamiltonian is the one in Eq. 3.4.6. In this case, it is favorable for the system also to maximize the projection of $\hat{S}_z$, and in this case $|0\rangle$ given in Eq. 3.6.2 is the *only* ground state, with energy

$$E_0(B) = -\frac{S^2}{2} \sum_{ij} J_{ij} - g\mu_B BNS. \tag{3.6.6}$$

In this case, the symmetry is not spontaneously broken, since it is already broken at the Hamiltonian level by the applied magnetic field.

Check Points

- What is the ground state of the ferromagnetic Heisenberg Hamiltonian?
- How do you prove it is the ground state?
- What happens in the presence of an external magnetic field to the degeneracy of the state?

3.7 Ground State of the Antiferromagnetic Heisenberg Hamiltonian

Aside from the ferromagnetic case treated in the previous section, finding the ground state of the Heisenberg Hamiltonian is in general difficult and has to be studied case by case. The ground state will depend on the nature of the interactions (short or long range, sign, anisotropy), from the lattice structure, and from the dimensionality of the system. To illustrate this difficulty, we discuss here briefly the antiferromagnetic Heisenberg model on a bipartite lattice [9]. In this case $J_{ij} \leq 0 \,\forall i,\,j$ and the lattice can be subdivided into two sublattices A and B, such that J_{ij} is finite only when i and j belong to two different sublattices. The simplest example is that of a square lattice with nearest-neighbor interactions, where all nearest neighbors of A belong to the B sublattice, and vice versa, see Fig. 3.2.

A guess of the ground state based on on the classical model is the so-called *Néel state*: the two sublattices are fully polarized, but in opposite directions

$$|0?\rangle_{\mathrm{AF}} = \prod_{i \in A} |S, S\rangle_i \prod_{j \in B} |S, -S\rangle_j \,.$$

It is easy to see, however, that this state is not an eigenstate of the Heisenberg Hamiltonian. Using the representation of $\hat{H}$ in terms of ladder operators, see Eq. 3.6.5, we see that

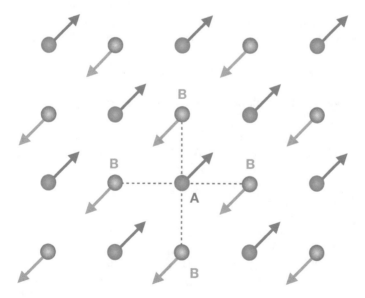

Fig. 3.2 Antiferromagnetic Heisenberg model on a bipartite square lattice, the dotted lines indicate nearest-neighbor interactions. Example of a Néel state

$$\hat{H}|0?\rangle_{\mathrm{AF}} = -\frac{1}{2}\sum_{i\in A}\sum_{i\in B} J_{ij} \left[\frac{1}{2}\left(\hat{S}_i^+ \hat{S}_j^- |0?\rangle_{\mathrm{AF}} + \hat{S}_i^- \hat{S}_j^+ |0?\rangle_{\mathrm{AF}} \right) + \hat{S}_i^z \hat{S}_j^z |0?\rangle_{\mathrm{AF}} \right].$$

The first term is simply zero, since the sublattices A and B are fully polarized in "the right way" with respect to the operators. The last term is simply proportional to $|0?\rangle_{\mathrm{AF}}$. However, for the second term

$$\hat{S}_i^- \hat{S}_j^+ |0?\rangle_{\mathrm{AF}} \propto |S,\rangle ... |S, -S+1\rangle_j ... |S, S-1\rangle_i ... |S, m_N\rangle$$

which shows that $|0?\rangle_{\mathrm{AF}}$ cannot be an eigenstate. In general, one can prove that the true ground state $|0\rangle_{\mathrm{AF}}$ is nondegenerate and a singlet of total spin: $\hat{S}_{\mathrm{TOT}}|0\rangle_{\mathrm{AF}} = 0$. This is called the *Marshall's Theorem*. Which kind of singlet is the actual ground state is, however, not determined.

Check Points

- What is a Néel state?
- Why it is not the ground state of the antiferromagnetic Heisenberg Hamiltonian?

3.8   Ground State of the Classical Heisenberg Model

The classical Heisenberg model is obtained by replacing the spin operators $\hat{S}_i$ by simple vectors S_i with fixed length $|S_i| = S$. As we saw in previous sections, this kind of approximation is valid in the limit of large spin S. For the classical model, it is straightforward to find the ground state for translational invariant systems,

$$J_{ij} = J_{ji} = J(\mathbf{R}_i - \mathbf{R}_j), \tag{3.8.1}$$

where $\mathbf{R}_i$ indicates the points on a Bravais lattice. In Fourier space

$$\mathbf{S}_i = \frac{1}{\sqrt{N}}\sum_{\mathbf{k}} \mathbf{S}_{\mathbf{k}} e^{i\mathbf{k}\cdot\mathbf{R}_i} \tag{3.8.2}$$

$$\mathbf{S}_{\mathbf{k}} = \frac{1}{\sqrt{N}}\sum_{i} \mathbf{S}_i e^{-i\mathbf{k}\cdot\mathbf{R}_i}$$

and hence

$$H_{\mathrm{cl}} = -\frac{1}{2N}\sum_{\mathbf{R}_i \mathbf{R}_j} J(\mathbf{R}_i - \mathbf{R}_j) \sum_{\mathbf{k}\mathbf{k}'} \left(e^{i\mathbf{k}\cdot\mathbf{R}_i}\mathbf{S}_{\mathbf{k}} \cdot \mathbf{S}_{\mathbf{k}'} e^{i\mathbf{k}'\cdot\mathbf{R}_j} \right).$$

Defining $\Delta\mathbf{R} = \mathbf{R}_i - \mathbf{R}_j$ and using

$$\sum_{\mathbf{R}_i} e^{i\mathbf{k}\cdot\mathbf{R}_i} = N\delta_{\mathbf{k},0} \qquad (3.8.3)$$

being $\delta_{\mathbf{k},\mathbf{k}'}$ the Kronecker Delta, we obtain

$$
\begin{aligned}
H_{\mathrm{cl}} &= -\frac{1}{2N}\sum_{\mathbf{R}_i\,\Delta\mathbf{R}} J(\Delta\mathbf{R})\sum_{\mathbf{k}\mathbf{k}'}\left(e^{i\mathbf{k}\cdot\mathbf{R}_i}\mathbf{S}_{\mathbf{k}}\cdot\mathbf{S}_{\mathbf{k}'}e^{i\mathbf{k}'\cdot\mathbf{R}_i}e^{-i\mathbf{k}'\cdot\Delta\mathbf{R}}\right)\\
&= -\frac{1}{2N}\sum_{\Delta\mathbf{R}} J(\Delta\mathbf{R})\sum_{\mathbf{k}\mathbf{k}'} N\delta_{\mathbf{k}+\mathbf{k}',0}\mathbf{S}_{\mathbf{k}}\cdot\mathbf{S}_{\mathbf{k}'}e^{-i\mathbf{k}'\cdot\Delta\mathbf{R}}\\
&= -\frac{1}{2}\sum_{\Delta\mathbf{R}} J(\Delta\mathbf{R})\sum_{\mathbf{k}}\mathbf{S}_{\mathbf{k}}\cdot\mathbf{S}_{-\mathbf{k}}e^{i\mathbf{k}\cdot\Delta\mathbf{R}}
\end{aligned}
$$

and defining

$$J(\mathbf{k}) = \sum_i J(\Delta\mathbf{R})e^{i\mathbf{k}\cdot\Delta\mathbf{R}} \qquad (3.8.4)$$

we obtain

$$H_{\mathrm{cl}} = -\frac{1}{2}\sum_{\mathbf{k}} J(\mathbf{k})\mathbf{S}_{\mathbf{k}}\cdot\mathbf{S}_{-\mathbf{k}}. \qquad (3.8.5)$$

Imposing the constraint $\mathbf{S}^2 = S^2$, one can show that the minimum of the energy given by Eq. 3.8.5 is given by setting $\mathbf{k} = \mathbf{Q}$, where $\mathbf{Q}$ determines the global maximum of $J(\mathbf{Q})$ [10]. One obtains the equilibrium configuration

$$\mathbf{S}_i = S\left(\cos(\mathbf{Q}.\mathbf{R}_i), \sin(\mathbf{Q}.\mathbf{R}_i), 0\right)$$

which is in general a planar helical state. Since $\mathbf{Q}$ is determined by the maximum of $J(\mathbf{Q})$, the order is not necessarily commensurate with the lattice, unless this maximum coincides with a high symmetry point in the Brillouin zone. Special cases are $\mathbf{Q} = 0$, where we recover ferromagnetic order, or $\mathbf{Q}$ taking a value at the edge of the Brillouin zone, in which case the order is antiferromagnetic. Applying a magnetic field in the z direction tilts the magnetic order out of plane.

Check Points

- What is the classical Heisenberg model and how does one in general find its ground state?

3.9   Dipole–Dipole Interactions

In the Heisenberg Hamiltonian, we have not included the dipole–dipole interaction term

$$\hat{H}_{d-d} = -\frac{\mu_0}{4\pi} \frac{(\mu_B g)^2}{2} \sum_{i \neq j} \frac{1}{|\mathbf{r}_i - \mathbf{r}_j|^3} \left[\frac{3\hat{\mathbf{S}}_i \cdot (\mathbf{r}_i - \mathbf{r}_j)\,\hat{\mathbf{S}}_j \cdot (\mathbf{r}_i - \mathbf{r}_j)}{|\mathbf{r}_i - \mathbf{r}_j|^2} - \hat{\mathbf{S}}_i \cdot \hat{\mathbf{S}}_j \right].$$

$$(3.9.1)$$

We argued this term is too small to justify magnetic ordering at the observed temperatures, and we found out that the exchange interaction is orders of magnitude larger. However, the exchange interaction depends on the overlap of orbitals, and as such is generally short ranged: it decays exponentially with distance, at least for insulators. From Eq. 3.9.1, we see that the dipole–dipole interaction decays instead algebraically, as $1/r^3$, and that means that it will be dominant at large distances. This gives rise, in large samples, to the formation of domains. In intermediate-size samples, usually micrometer sized, the competition between dipole–dipole interactions and exchange interactions leads to textured ground states. Being long ranged, the dipolar interactions are sensitive to the boundaries of the sample, where the spins try to align with the boundary, so as to minimize the "magnetic monopoles" at the surface as introduced in Chap. 1. Dipole–dipole interaction terms can be included in the Heisenberg Hamiltonian in the long-wavelength limit as demagnetization fields.

Chapter 4
Spin Waves and Magnons

In the last chapter, we dealt with the ground state of the Heisenberg model for a few solvable examples. The ground state is the only eigenstate that is occupied at strictly zero temperature. The excitations on top of this ground state will determine the behavior of the system at low temperatures. These are collective excitations of the magnetic system, and their quanta are denominated magnons [11]. One can draw an analogy to another physical phenomenon which is perhaps more intuitive: that of mechanical vibrations. There, all atoms participate in the collective mechanical vibration, and we call phonons the respective quanta. In the next sections, we will study the magnetic elementary excitations in more detail. We will focus on the simplest case of the Heisenberg ferromagnet, but the concepts are general.

4.1 Excitations of the Heisenberg Ferromagnet

For the Heisenberg ferromagnet at $T = 0$, we found that all spins are aligned with and the magnetization is given by the saturation magnetization

$$M_s = g\mu_B \frac{N}{V} S.$$

At $T \neq 0$, some spins will "flip" or, more generally, decrease by one unit, which in turn will decrease the magnetization from its saturation value. An obvious candidate for an excited state is therefore

$$|i\rangle = |S, S\rangle_1 \ldots |S, S - 1\rangle_i \ldots |S, S\rangle_N \tag{4.1.1}$$

which can be obtained from the ground state Eq. 3.6.2 as

© The Author(s), under exclusive license to Springer Nature Switzerland AG 2019
S. Viola Kusminskiy, *Quantum Magnetism, Spin Waves, and Optical Cavities*,
SpringerBriefs in Physics, https://doi.org/10.1007/978-3-030-13345-0_4

$$|i\rangle = \frac{1}{\sqrt{2S}} \hat{S}_i^- |0\rangle . \tag{4.1.2}$$

One can easily see that $|i\rangle$ is an eigenstate of the $\hat{S}^z$ operator

$$\hat{S}_i^z |i\rangle = (S - 1)|i\rangle$$
$$\hat{S}_j^z |j\rangle = S|j\rangle \quad \forall \quad j \neq i .$$

However, $\hat{S}_i^+ |i\rangle \neq 0$, instead one obtains

$$\hat{S}_j^- \hat{S}_i^+ |i\rangle = 2S|j\rangle .$$

Therefore, the Heisenberg Hamiltonian 3.6.5 shifts the site of the "flipped" spin $i \to j$, and $|i\rangle$ as given in Eq. 4.1.1 is not an eigenstate of 3.6.5. It is also not a good approximation to an excited eigenstate, since flipping a spin in such a manner has a very high energy cost, of the order of the exchange interaction. For example, for nearest-neighbor interaction with exchange constant J, flipping a spin has an energy cost of $\Delta E \sim zJS$, where z is the *coordination number* (that is, z is the number of nearest neighbors, e.g., for a square lattice $z = 4$). We have already seen that the exchange constant is of the order of the critical temperature, quite a high energy for regular ferromagnets.

One can prove that an actual eigenstate of 3.6.5 is given by

$$|\mathbf{k}\rangle = \frac{1}{\sqrt{N}} \sum_{\mathbf{R}_i} e^{-i\mathbf{R}_i \cdot \mathbf{k}} |i\rangle , \tag{4.1.3}$$

where $\mathbf{R}_i$ are the lattice sites. We see therefore that an eigenstate is formed by distributing the "flipped" spin over all sites, and therefore it is a *collective excitation* of the system. This collective excitation has well-defined momentum $\hbar\mathbf{k}$ (up to a reciprocal lattice vector) and energy $\hbar\omega(\mathbf{k})$ and we call it a *quasiparticle*. In particular, for magnetic systems these quasiparticles are denominated *magnons*. In the following, we will show that $|\mathbf{k}\rangle$ is an eigenstate of 3.6.5 and will calculate its *dispersion relation* $\hbar\omega(\mathbf{k})$.

In order to do this, we first perform a Fourier transform of the Hamiltonian 3.6.5 by defining the spin operators in momentum space (in analogy to the classical case, see Eq. 3.8.4)

$$\hat{S}_{\mathbf{k}}^\alpha = \frac{1}{\sqrt{N}} \sum_{\mathbf{R}_i} e^{-i\mathbf{k} \cdot \mathbf{R}_i} \hat{S}_i^\alpha \tag{4.1.4}$$

$$\hat{S}_i^\alpha = \frac{1}{\sqrt{N}} \sum_{\mathbf{k}} \hat{S}_{\mathbf{k}}^\alpha e^{i\mathbf{k} \cdot \mathbf{R}_i} ,$$

where $\alpha = x$, y, z, $\pm$. Note that

$$\left(\hat{S}_{\mathbf{k}}^+\right)^\dagger = \hat{S}_{-\mathbf{k}}^-$$

$$\left(\hat{S}_{\mathbf{k}}^z\right)^\dagger = \hat{S}_{-\mathbf{k}}^z$$

and one can verify that the commutators now read

$$\left[\hat{S}_{\mathbf{k}_1}^+, \hat{S}_{\mathbf{k}_2}^-\right] = 2\hat{S}_{\mathbf{k}_1+\mathbf{k}_2}^z \tag{4.1.5}$$

$$\left[\hat{S}_{\mathbf{k}_1}^z, \hat{S}_{\mathbf{k}_2}^\pm\right] = \pm\hat{S}_{\mathbf{k}_1+\mathbf{k}_2}^\pm .$$

Using that $J_{ij} = J_{ji}$ and that spin operators at different sites commute, we rewrite slightly the Heisenberg Hamiltonian 3.6.5 as

$$\hat{H} = -\frac{1}{2}\sum_{ij} J_{ij}\left[\hat{S}_i^+\hat{S}_j^- + \hat{S}_i^z\hat{S}_j^z\right] . \tag{4.1.6}$$

From Eqs. 4.1.4 and using again Eq. 3.8.1, we obtain

$$\hat{H} = -\frac{1}{2N}\sum_{\mathbf{R}_i\mathbf{R}_j} J(\mathbf{R}_i - \mathbf{R}_j)\sum_{\mathbf{k}\mathbf{k}'}\left(\hat{S}_{\mathbf{k}}^+ e^{i\mathbf{k}\cdot\mathbf{R}_i}\hat{S}_{\mathbf{k}'}^- e^{i\mathbf{k}'\cdot\mathbf{R}_j} + \hat{S}_{\mathbf{k}}^z e^{i\mathbf{k}\cdot\mathbf{R}_i}\hat{S}_{\mathbf{k}'}^z e^{i\mathbf{k}'\cdot\mathbf{R}_j}\right)$$
$$- \frac{g\mu_B}{\sqrt{N}}B_0\sum_{\mathbf{R}_i}\sum_{\mathbf{k}}\hat{S}_{\mathbf{k}}^z e^{i\mathbf{k}\cdot\mathbf{R}_i},$$

where we added an external magnetic field. We use $\Delta\mathbf{R} = \mathbf{R}_i - \mathbf{R}_j$ and Eq. 3.8.3 to write

$$\hat{H} = -\frac{1}{2N}\sum_{\Delta\mathbf{R}} J(\Delta\mathbf{R})\sum_{\mathbf{R}_i}\sum_{\mathbf{k}\mathbf{k}'} e^{i(\mathbf{k}+\mathbf{k}')\cdot\mathbf{R}_i}\left(\hat{S}_{\mathbf{k}}^+\hat{S}_{\mathbf{k}'}^- + \hat{S}_{\mathbf{k}}^z\hat{S}_{\mathbf{k}'}^z\right)e^{-i\mathbf{k}'\cdot\Delta\mathbf{R}}$$
$$- g\mu_B\sqrt{N}B_0\sum_{\mathbf{k}}\hat{S}_{\mathbf{k}}^z\delta_{\mathbf{k},0}$$
$$= -\frac{1}{2N}\sum_{\Delta\mathbf{R}} J(\Delta\mathbf{R})N\sum_{\mathbf{k}\mathbf{k}'}\delta_{\mathbf{k}+\mathbf{k}',0}\left(\hat{S}_{\mathbf{k}}^+\hat{S}_{\mathbf{k}'}^- + \hat{S}_{\mathbf{k}}^z\hat{S}_{\mathbf{k}'}^z\right)e^{-i\mathbf{k}'\cdot\Delta\mathbf{R}} - g\mu_B\sqrt{N}B_0\hat{S}_0^z$$
$$= -\frac{1}{2}\sum_{\Delta\mathbf{R}} J(\Delta\mathbf{R})\sum_{\mathbf{k}}\left(\hat{S}_{\mathbf{k}}^+\hat{S}_{-\mathbf{k}}^- + \hat{S}_{\mathbf{k}}^z\hat{S}_{-\mathbf{k}}^z\right)e^{i\mathbf{k}\cdot\Delta\mathbf{R}} - g\mu_B\sqrt{N}B_0\hat{S}_0^z .$$

Using the definition for $J(\mathbf{k})$ given in Eq. 3.8.4, we finally obtain an expression for the Heisenberg Hamiltonian in momentum space

$$\hat{H} = -\frac{1}{2} \sum_{\mathbf{k}} J(\mathbf{k}) \left(\hat{S}_{\mathbf{k}}^+ \hat{S}_{-\mathbf{k}}^- + \hat{S}_{\mathbf{k}}^z \hat{S}_{-\mathbf{k}}^z \right) - g\mu_B \sqrt{N} B_0 \hat{S}_0^z . \qquad (4.1.7)$$

First, we check the action of $\hat{H}$ given in Eq. 4.1.7 on the ground state $|0\rangle$. For that we note

$$\hat{S}_{\mathbf{k}}^z |0\rangle = \frac{1}{\sqrt{N}} \sum_{\mathbf{R}_i} e^{-i\mathbf{k}\cdot\mathbf{R}_i} \hat{S}_i^z |0\rangle = \frac{S}{\sqrt{N}} \sum_{\mathbf{R}_i} e^{-i\mathbf{k}\cdot\mathbf{R}_i} |0\rangle = S\sqrt{N}\delta_{\mathbf{k},0} |0\rangle$$

$$\hat{S}_i^+ |0\rangle = 0 \quad \Rightarrow \quad \hat{S}_{\mathbf{k}}^+ |0\rangle = 0$$

and examine the action of each term in Eq. 4.1.7 on $|0\rangle$. Using the commutators in Eq. 4.1.5, for the first term in the sum, we obtain

$$-\frac{1}{2} \sum_{\mathbf{k}} J(\mathbf{k}) \hat{S}_{\mathbf{k}}^+ \hat{S}_{-\mathbf{k}}^- |0\rangle = -\frac{1}{2} \sum_{\mathbf{k}} J(\mathbf{k}) \left(\hat{S}_{-\mathbf{k}}^- \hat{S}_{\mathbf{k}}^+ + 2\hat{S}_0^z \right) |0\rangle$$

$$= -\frac{1}{2} \sum_{\mathbf{k}} J(\mathbf{k}) \left(2\hat{S}_0^z \right) |0\rangle = -S\sqrt{N} \sum_{\mathbf{k}} J(\mathbf{k}) |0\rangle = 0,$$

where the last equality stems from

$$\sum_{\mathbf{k}} J(\mathbf{k}) = \sum_{\mathbf{k}} \sum_{\Delta\mathbf{R}} J(\Delta\mathbf{R}) e^{i\mathbf{k}\cdot\Delta\mathbf{R}} = \sum_{\Delta\mathbf{R}} J(\Delta\mathbf{R}) N\delta_{\Delta\mathbf{R},0} = N J(\Delta\mathbf{R}=0) = 0 .$$

For the second term in Eq. 4.1.7

$$-\frac{1}{2} \sum_{\mathbf{k}} J(\mathbf{k}) \hat{S}_{\mathbf{k}}^z \hat{S}_{-\mathbf{k}}^z |0\rangle = -\frac{1}{2} \sqrt{N} S \sum_{\mathbf{k}} J(\mathbf{k}) \hat{S}_{\mathbf{k}}^z \delta_{-\mathbf{k},0} |0\rangle$$

$$= -\frac{1}{2} \sqrt{N} S J(\mathbf{k}=0) \hat{S}_0^z |0\rangle$$

$$= -\frac{1}{2} N S^2 J(\mathbf{k}=0) |0\rangle .$$

We now note that

$$N J(\mathbf{k}=0) = N \sum_{\Delta\mathbf{R}} J(\Delta\mathbf{R}) = \sum_{\mathbf{R}_i \Delta\mathbf{R}} J(\Delta\mathbf{R}) = \sum_{\mathbf{R}_i \mathbf{R}_j} J(\mathbf{R}_i - \mathbf{R}_j) = \sum_{ij} J_{ij}$$

and therefore

$$-\frac{1}{2} \sum_{\mathbf{k}} J(\mathbf{k}) \hat{S}_{\mathbf{k}}^z \hat{S}_{-\mathbf{k}}^z |0\rangle = -\frac{S^2}{2} \sum_{ij} J_{ij} |0\rangle .$$

For the third term in Eq. 4.1.7, we obtain simply

$$-g\mu_B\sqrt{N}B_0\hat{S}_0^z|0\rangle = -g\mu_B N B_0 S|0\rangle .$$

Putting all together, we obtain $\hat{H}|0\rangle = E_0|0\rangle$ with

$$E_0(B_0) = -\frac{S^2}{2}\sum_{ij}J_{ij} - g\mu_B B_0 N S$$

which coincides with our result in Eq. 3.6.6 as expected.

We now want to show that $|\mathbf{k}\rangle$ given in Eq. 4.1.3 is a eigenstate of 4.1.7. Using 4.1.2, we can write $|\mathbf{k}\rangle$ in terms of the ladder operators in momentum space,

$$|\mathbf{k}\rangle = \frac{1}{\sqrt{2S}}\hat{S}_\mathbf{k}^-|0\rangle , \tag{4.1.8}$$

hence it is enough to show that $\hat{S}_\mathbf{k}^-|0\rangle$ is an eigenstate [3]. Writing

$$\hat{H}\hat{S}_\mathbf{k}^-|0\rangle = \left(\hat{S}_\mathbf{k}^-\hat{H} + \left[\hat{H},\hat{S}_\mathbf{k}^-\right]\right)|0\rangle = \left(\hat{S}_\mathbf{k}^- E_0 + \left[\hat{H},\hat{S}_\mathbf{k}^-\right]\right)|0\rangle$$
$$= E_0\hat{S}_\mathbf{k}^-|0\rangle + \left[\hat{H},\hat{S}_\mathbf{k}^-\right]|0\rangle ,$$

we see that this amounts to showing that $\left[\hat{H},\hat{S}_\mathbf{k}^-\right]|0\rangle \propto \hat{S}_\mathbf{k}^-|0\rangle$. Using similar manipulations as above, and taking into account that $J(\mathbf{k}) = J(-\mathbf{k})$, one obtains

$$\left[\hat{H},\hat{S}_\mathbf{k}^-\right]|0\rangle = [g\mu_B B_0 - S(J(\mathbf{k}) - J(\mathbf{k}=0))]\,\hat{S}_\mathbf{k}^-|0\rangle \tag{4.1.9}$$

and therefore $|\mathbf{k}\rangle$ is an eigenstate of 4.1.7 with eigenvalue

$$E(\mathbf{k}) = E_0(B_0) + g\mu_B B_0 - S(J(\mathbf{k}) - J(\mathbf{k}=0)) . \tag{4.1.10}$$

The energy on top of the ground state is the *excitation energy*

$$\hbar\omega(\mathbf{k}) = g\mu_B B_0 - S[J(\mathbf{k}) - J(\mathbf{k}=0)], \tag{4.1.11}$$

which is simply the *energy of one magnon*—the quasiparticle energy is always defined with respect to the ground state energy; in field theory, it is common to call the ground state of a system the "vacuum". Equation 4.1.11 is also called the *dispersion relation*, since it gives the dependence of the energy with the wave vector. Note that for $B_0 = 0$, $\hbar\omega(0) = 0$ and therefore any infinitesimal temperature will cause excitations with $\mathbf{k} = 0$. This "zero mode" is an example of a *Goldstone mode*, which is always present when there is spontaneous symmetry breaking in the system. In general, Goldstone modes are massless bosons quasiparticles which appear in systems where there is a spontaneously broken continuous symmetry.

If we compare with the ground state energy $E_0(B_0)$, in particular, the Zeeman term, we see that the magnetic moment of the system has been modified by one unit. We can therefore conclude that a *magnon has spin 1*, and therefore it is a *bosonic* quasiparticle. The expectation value of a local spin operator $\hat{S}_i^z$ with respect to the one-magnon state $|\mathbf{k}\rangle$ can be shown to be

$$\langle \mathbf{k}|\hat{S}_i^z|\mathbf{k}\rangle = S - \frac{1}{N} = \langle 0|\hat{S}_i^z|0\rangle - \frac{1}{N} \quad \forall \ i, \mathbf{k} \tag{4.1.12}$$

which shows that the spin reduction is indeed of one unit and it is distributed uniformly over all sites $\mathbf{R}_i$. Semiclassically, one can picture each spin in the lattice precessing around the z-axis with a projection of $\hbar(S - 1/N)$. However, except for $\mathbf{k} = 0$, the spins do not precess in phase but instead they have a phase difference of $e^{-i\mathbf{k}\cdot(\mathbf{R}_i-\mathbf{R}_j)}$, forming a *spin wave*. This semiclassical picture can be better understood if we look at the equations of motion for the spins, which we do in the following section.

We finish this section by stating that our intuition fails again if we want to construct two-magnon states. The obvious choice

$$|\mathbf{k}, \mathbf{k}'\rangle \propto \hat{S}_{\mathbf{k}}^- \hat{S}_{\mathbf{k}'}^-|0\rangle \tag{4.1.13}$$

is actually not an eigenstate of the Heisenberg Hamiltonian. This is due to magnon–magnon interactions present in the Hamiltonian, which are not taken into account in a simple product state as 4.1.13. The same of course holds for multiple-magnon states. Therefore, when having multiple magnons excited in a system, they can interact and magnon states will decay due to magnon–magnon interactions. These excited states therefore have a certain *lifetime*. Besides magnon–magnon interactions, scattering with impurities or phonons in a material will determine the lifetime of magnon states.

Check Points

- What is a magnon?
- How do you obtain the energy dispersion of a magnon?

4.2   Equation of Motion Approach

In the Heisenberg picture, the time dependence is included in the operators. Instead of the Schrödinger equation for the state vectors, we write the *Heisenberg equation of motion* for the spin operators

$$\hbar\frac{d\hat{\mathbf{S}}_i}{dt} = i[\hat{H}, \hat{\mathbf{S}}_i]. \tag{4.2.1}$$

Using the Heisenberg Hamiltonian Eq. 3.4.6 and the commutation relations for the spin operators, it is straightforward to show

$$\hbar \frac{d\hat{\mathbf{S}}_i}{dt} = -\left(\sum_j J_{ij} \hat{\mathbf{S}}_j + g\mu_B \mathbf{B_0} \right) \times \hat{\mathbf{S}}_i . \tag{4.2.2}$$

This equation is exact. We can compare it, however, with the classical equation of motion for the angular momentum given in Eq. 1.5.2, and we see that, by recovering the units of the spin and absorbing the Planck constant into the definition of the exchange constant, Eq. 4.2.2 can be directly translated into a classical equation of motion by taking the expectation values of the spin operators, and in particular

$$\mathbf{B}_{eff} = \mathbf{B_0} + \frac{1}{g\mu_B} \langle \sum_j J_{ij} \hat{\mathbf{S}}_j \rangle$$

is the effective magnetic field including the Weiss molecular fields.

We turn now to a different approximation, in which we retain for now the operator character of $\hat{\mathbf{S}}_i$. We are, as in the previous section, interested in the low-energy excitations of the system on top of the ground state. Since the ground state is fully polarized, we expect the projection $\hat{S}_i^z$ to remain almost constant and close to S. From Eq. 4.1.12, we see this is valid as long as $NS \gg 1$, that is, the total spin number is much larger than the number of excitations in the system. From Eq. 4.2.2, we obtain

$$\hbar \frac{d\hat{S}_i^x}{dt} \approx -S \sum_j J_{ij} \left(\hat{S}_j^y - \hat{S}_i^y \right) + g\mu_B B_0 \hat{S}_i^y \tag{4.2.3}$$

$$\hbar \frac{d\hat{S}_i^y}{dt} \approx -S \sum_j J_{ij} \left(\hat{S}_i^x - \hat{S}_j^x \right) - g\mu_B B_0 \hat{S}_i^x$$

$$\hbar \frac{d\hat{S}_i^z}{dt} \approx 0 .$$

These equations are decoupled for the ladder operators

$$\hbar \frac{d\hat{S}_i^{\pm}}{dt} = \mp i \left[S \sum_j J_{ij} \left(\hat{S}_i^{\pm} - \hat{S}_j^{\pm} \right) + g\mu_B B_0 \hat{S}_i^{\pm} \right] , \tag{4.2.4}$$

where we write the equal sign in the understanding that the equation is valid in the limit established for Eqs. 4.2.3.

Equation 4.2.4 still couple spin operators at different sites. To decouple them, we go once more to the Fourier representation. Using Eq. 4.1.4, for the lowering operator, one obtains

$$\hbar\frac{d\hat{S}_{\mathbf{k}}^{-}}{dt} = iS\left[J(\mathbf{k}=\mathbf{0}) - J(\mathbf{k})\right]\hat{S}_{\mathbf{k}}^{-} + ig\mu_{B}B_{0}\hat{S}_{\mathbf{k}}^{-}. \tag{4.2.5}$$

Hence, we see that the equation of motion for each wave vector $\mathbf{k}$ is decoupled from the rest, in the spirit of a normal mode's decomposition. Equation 4.2.5 is easily solved by

$$\hat{S}_{\mathbf{k}}^{-} = \hat{M}_{\mathbf{k}}e^{i\omega(\mathbf{k})t + i\alpha_{\mathbf{k}}} \tag{4.2.6}$$

with the time dependence given entirely by the exponential term, and $\omega(\mathbf{k})$ is given by Eq. 4.1.11 ($\alpha_{\mathbf{k}}$ is an arbitrary phase). We have therefore re-derived the dispersion relation obtained in the previous section for the one-magnon state.

In the semiclassical picture and considering only one excited mode, for the real-space components of the spin, we obtain

$$S_i^x = \frac{M_{\mathbf{k}}}{\sqrt{N}}\cos\left(\mathbf{k}\cdot\mathbf{R}_i + \omega(\mathbf{k})t\right)$$

$$S_i^y = \frac{M_{\mathbf{k}}}{\sqrt{N}}\sin\left(\mathbf{k}\cdot\mathbf{R}_i + \omega(\mathbf{k})t\right)$$

$$S_i^z = S$$

which are the components of a plane wave with frequency $\omega(\mathbf{k})$.

Coming back to the Heisenberg equation of motion for $\hat{S}_{\mathbf{k}}^{-}$

$$\hbar\frac{d\hat{S}_{\mathbf{k}}^{-}}{dt} = i[\hat{H}, \hat{\mathbf{S}}_{\mathbf{k}}^{-}]$$

from Eq. 4.2.6, we can write

$$\hbar\omega(\mathbf{k})\hat{S}_{\mathbf{k}}^{-}|0\rangle = [\hat{H}, \hat{\mathbf{S}}_{\mathbf{k}}^{-}]|0\rangle = \hat{H}\hat{S}_{\mathbf{k}}^{-}|0\rangle - \hat{S}_{\mathbf{k}}^{-}E_0|0\rangle$$

and therefore we recover

$$\hat{H}\hat{S}_{\mathbf{k}}^{-}|0\rangle = [E_0 - \hbar\omega(\mathbf{k})]\,\hat{S}_{\mathbf{k}}^{-}|0\rangle$$

as in the exact result.

As an example, we give the dispersion relation for spins on a cubic lattice with nearest-neighbor interactions. For this case,

$$J(\mathbf{k}) = 2J\left(\cos(k_x a) + \cos(k_y a) + \cos(k_z a)\right)$$

and therefore

$$\hbar\omega(\mathbf{k}) = 2JS\left(3 - \cos(k_x a) + \cos(k_y a) + \cos(k_z a)\right) + g\mu_B B_0.$$

For small $\mathbf{k}$, we obtain a quadratic dispersion

$$\hbar\omega(\mathbf{k}) \approx JSk^2a^2 + g\mu_B B_0, \tag{4.2.7}$$

which is gapped as long as $B_0 \neq 0$. Note that this dispersion is different from the usual acoustic phonon dispersion in solids, which is linear in k. The dispersion 4.2.7 is actually similar to that encountered for *flexural* phonon modes in materials with reduced dimensionality: for example, the out-of-plane phonon modes of a graphene membrane. The k^2 dispersion is a signature of the rotational symmetry of the problem.

Check Points

- Derive the Heisenberg equation of motion for the spins from the Heisenberg Hamiltonian.
- Relate the concept of magnon from the previous section, with the semiclassical picture given by the equations of motion.

4.3 Holstein–Primakoff Transformation

Due to their algebra (that is, their commutation relations), angular momentum operators are difficult to treat in an interacting theory. There are, however, transformations which write the angular momentum operators in terms of second-quantization creation and annihilation operators, either fermionic or bosonic. The idea of these transformations is to simplify the commutation rules, so that one can use well-known methods of second quantization. The price to pay is that the transformations are nonlinear. In this section, we go over one of these transformations which is widely used, the Holstein–Primakoff transformation. Within this transformation, the angular momentum operators are written as nonlinear functions of bosonic creation and annihilation operators, that is, a collection of harmonic oscillators.

Before writing the transformation explicitly, we remind briefly the properties of *creation* ($\hat{a}_i^\dagger$) and *annihilation* ($\hat{a}_i$) harmonic oscillator operators. The subscript i indicates the lattice site, that is, we have a harmonic oscillator at every site on the lattice. The commutation relations for these bosonic operators are

$$\left[\hat{a}_i, \hat{a}_j^\dagger\right] = \delta_{ij} \tag{4.3.1}$$

$$\left[\hat{a}_i, \hat{a}_j\right] = \left[\hat{a}_i^\dagger, \hat{a}_j^\dagger\right] = 0$$

The Hamiltonian of a single oscillator reads

$$H_{\text{osc}} = \hbar\omega_i \left(\hat{a}_i^\dagger \hat{a}_i + \frac{1}{2}\right), \tag{4.3.2}$$

where ω_i is the frequency of oscillator i. If the oscillators are independent, the total Hamiltonian is simply the sum over the respective Hamiltonians. Usually, however, the original operators are not independent, but if the Hamiltonian is *quadratic* in these, one can find a linear transformation which diagonalizes the Hamiltonian, so that in the new basis the Hamiltonian is a sum of harmonic oscillators Eq. 4.3.2. A state $|n_i\rangle$ with n_i "particles" at site i can be constructed from the vacuum $|0_i\rangle$ by applying $\hat{a}_i^\dagger$ (below we will identify n_i with the number of flipped spins at site i). In general, we have

$$\hat{a}_i|0_i\rangle = 0 \qquad\qquad (4.3.3)$$
$$\hat{a}_i^\dagger|n_i\rangle = \sqrt{n_i + 1}|n_i + 1\rangle$$
$$\hat{a}_i|n_i\rangle = \sqrt{n_i}|n_i - 1\rangle$$
$$\hat{a}_i^\dagger\hat{a}_i|n_i\rangle = n_i|n_i\rangle = \hat{n}_i|n_i\rangle,$$

where $\hat{n}_i = \hat{a}_i^\dagger\hat{a}_i$ is the *number operator* at site i. Therefore, $|n_i\rangle$ is an eigenstate of H_{osc} with energy $E(n_i) = \hbar\omega_i (n_i + 1/2)$, with $n_i = 0, 1, 2, \dots$.

Comparing Eqs. 4.3.3 with Eq. 3.6.4 and $\hat{S}_i^z$

$$\hat{S}_i^\pm|S, m_i\rangle = \sqrt{(S \mp m_i)(S + 1 \pm m_i)}|S, m_i \pm 1\rangle$$
$$\hat{S}_i^z|S, m_i\rangle = m_i|S, m_i\rangle,$$

we see that the creation and annihilation bosonic operators act in a similar way to the spin ladder operators, whereas the number operator is diagonal in this basis just as $\hat{S}_i^z$. However, we cannot replace simply $\hat{S}_i^\pm$ by $\hat{a}_i^\dagger$, $\hat{a}_i$ since this would not satisfy the commutation relations for the spin. This is accounted for by using a nonlinear transformation, the *Holstein–Primakoff transformation*

$$\hat{S}_i^+ = \sqrt{2S}\sqrt{1 - \frac{\hat{a}_i^\dagger\hat{a}_i}{2S}}\,\hat{a}_i$$
$$\hat{S}_i^- = \sqrt{2S}\hat{a}_i^\dagger\sqrt{1 - \frac{\hat{a}_i^\dagger\hat{a}_i}{2S}}$$
$$\hat{S}_i^z = \left(S - \hat{a}_i^\dagger\hat{a}_i\right). \qquad\qquad (4.3.4)$$

As we anticipated, in this case, the number operator counts the number of flipped spins, which we can see from the last equality in Eqs. 4.3.4

$$\hat{S}_i^z|n_i\rangle = \left(S - \hat{n}_i\right)|n_i\rangle = (S - n_i)|n_i\rangle \equiv m_i|n_i\rangle.$$

For $n_i = 0$, the spin is fully polarized and $m_i = S$, and hence the Holstein–Primakoff vacuum $|n_i = 0\rangle$ corresponds to the fully polarized spin state. We see, however, that the spectrum of $\hat{n}_i$ is constrained, due to the square root in Eqs. 4.3.4 we must impose

$n_i = 0, 1, ...2S$. The maximum value of n_i corresponds to the spin "fully flipped", $m_i = -S$. We observe that

$$\hat{S}_i^- |m_i = -S\rangle = \sqrt{2S}\hat{a}_i^\dagger\sqrt{1 - \frac{\hat{n}_i}{2S}}|n_i = 2S\rangle = \sqrt{2S}\hat{a}_i^\dagger\sqrt{1 - \frac{2S}{2S}}|n_i = 2S\rangle = 0$$

$$\hat{S}_i^+ |m_i = -S - 1\rangle = \sqrt{2S}\sqrt{1 - \frac{\hat{n}_i}{2S}}\hat{a}_i|n_i = 2S + 1\rangle$$

$$= \sqrt{2S}\sqrt{2S + 1}\sqrt{1 - \frac{\hat{n}_i}{2S}}|n_i = 2S\rangle = 0$$

and hence the ladder operators do not connect the physical subspace $n_i = 0, 1, ...2S$ with the unphysical one $n_i > 2S$.

Let us now define

$$\phi(\hat{n}_i) = \sqrt{2S}\sqrt{1 - \frac{\hat{n}_i}{2S}} \qquad (4.3.5)$$

and write the Heisenberg Hamiltonian Eq. 3.4.1 in terms of the Holstein–Primakoff operators, Eqs. 4.3.4. We find

$$\hat{H} = -\frac{NS^2 J_0}{2} + SJ_0 \sum_i \hat{n}_i - S\sum_{ij} J_{ij}\phi(\hat{n}_i)\hat{a}_i\hat{a}_j^\dagger\phi(\hat{n}_j) - \frac{1}{2}\sum_{ij} J_{ij}\hat{n}_i\hat{n}_j \quad (4.3.6)$$

with $J_0 = \sum_i J_{ij}$. We see that, due to Eq. 4.3.5, this Hamiltonian is not quadratic in the $\hat{a}_i^\dagger, \hat{a}_i$ operators and therefore we cannot write it as a sum of independent harmonic oscillators. We have, hence, transformed our original Heisenberg Hamiltonian of interacting spins into a Hamiltonian of interacting bosons.

Check Points

- Write the Holstein–Primakoff transformation.
- What is special about it?
- What is a magnon in this language?
- What is the meaning of the Heisenberg Hamiltonian in terms of the bosonic operators?

4.4 Spin-Wave Approximation

We will now proceed to reformulate our Hamiltonian into a noninteracting term (namely, noninteracting magnons), plus interaction terms. If we expand Eq. 4.3.5 as a series

$$\phi(\hat{n}_i) = 1 - \frac{\hat{n}_i}{4S} - \frac{\hat{n}_i^2}{32S^2} - \cdots \qquad (4.4.1)$$

we can write Eq. 4.3.6 also as a series

$$\hat{H} = -\frac{N S^2 J_0}{2} + \sum_{n=1}^{\infty} : \hat{H}_{2n} : \qquad (4.4.2)$$

with $: \hat{H}_{2n} :$ containing n creation and n annihilation operators in normal order (all $\hat{a}_i^\dagger$ to the left, all $\hat{a}_i$ to the right, e.g., for $n = 2$, $\hat{a}_i^\dagger \hat{a}_j^\dagger \hat{a}_m \hat{a}_l$). The terms with $n > 1$, that is, beyond quadratic, give rise to magnon–magnon interactions as we will see below. We will, however, first study the *spin-wave approximation*, where only the quadratic, noninteracting terms are kept in the expansion Eq. 4.4.2

$$: \hat{H}_{2n} := S J_0 \sum_i \hat{n}_i - S \sum_{ij} J_{ij} \hat{a}_i^\dagger \hat{a}_j . \qquad (4.4.3)$$

This truncation of the series is justified at low temperatures, where the number of excitations (total number of flipped spins) is small compared with the total number of spins NS. For that to hold, the average number of flipped spins per site has to be small $n_i \ll S$, and therefore we can approximate the square root in Eq. 4.3.5 to 1 and hence $\phi(\hat{n}_i) \approx \sqrt{2S}$. Within this approximation, the spin ladder operators are indeed approximated by simple harmonic oscillators, while the z component is kept at saturation

$$\hat{S}_i^+ \approx \sqrt{2S} \hat{a}_i \qquad (4.4.4)$$
$$\hat{S}_i^- \approx \sqrt{2S} \hat{a}_i^\dagger$$
$$\hat{S}_i^z \approx S ,$$

and it can be directly seen that the Heisenberg Hamiltonian Eq. 3.4.1 is quadratic in the $\hat{a}_i^\dagger$, $\hat{a}_i$ operators. This approximation is completely analogous to the one we performed when working with the equation of motion, Eq. 4.2.3. Note that, although the Hamiltonian is quadratic, it is not diagonal in i, j (see Eq. 4.4.3). Just as we did for the equations of motion, we need to go to Fourier space to obtain a diagonal Hamiltonian and therefore decoupled harmonic oscillators. In this case, we transform simply the bosonic operators

$$\hat{a}_{\mathbf{k}} = \frac{1}{\sqrt{N}} \sum_{\mathbf{R}_i} e^{-i\mathbf{k}\cdot\mathbf{R}_i} \hat{a}_i$$

$$\hat{a}_{\mathbf{k}}^\dagger = \frac{1}{\sqrt{N}} \sum_{\mathbf{R}_i} e^{i\mathbf{k}\cdot\mathbf{R}_i} \hat{a}_i^\dagger$$

and, within this approximation, we can show that the Heisenberg Hamiltonian reduces to

$$\hat{H}_{sw} = E(B_0) + \sum_{\mathbf{k}} \hbar\omega(\mathbf{k})\hat{a}_{\mathbf{k}}^{\dagger}\hat{a}_{\mathbf{k}} \qquad (4.4.5)$$

where we have added an external magnetic field for completeness, and $E(B_0)$ and $\omega(\mathbf{k})$ are given by Eqs. 4.1.10 and 4.1.11, respectively.

The Hamiltonian 4.4.5 describes a system of uncoupled harmonic oscillators. Its eigenstates are simply products of one-magnon states, that can be obtained from the vacuum by applying repeatedly $\hat{a}_{\mathbf{k}}^{\dagger}$

$$|\psi_{SW}\rangle = \prod_{\mathbf{k}} \left(\hat{a}_{\mathbf{k}}^{\dagger}\right)^{n_{\mathbf{k}}} |0\rangle, \qquad (4.4.6)$$

where $n_{\mathbf{k}}$ is the number of magnons with wavevector $\mathbf{k}$, eigenvalue of the number operator in Fourier representation $\hat{n}_{\mathbf{k}} = \hat{a}_{\mathbf{k}}^{\dagger}\hat{a}_{\mathbf{k}}$. We see that we can write the one-magnon state defined in Eq. 4.1.3 as $|\mathbf{k}\rangle = \hat{a}_{\mathbf{k}}^{\dagger}|0\rangle$. This state is an eigenstate of both the full Heisenberg Hamiltonian and of the noninteracting Hamiltonian 4.4.5. States with $n_{\mathbf{k}} > 1$ are, however, only eigenstates of 4.4.5.

Check Points

- What is the meaning of the spin-wave approximation in terms of the Holstein–Primakoff transformation?

4.5 Magnon–Magnon Interactions

We now proceed to investigate the higher order terms ($n > 1$) in Eq. 4.4.2. We consider for simplicity a Heisenberg Hamiltonian with nearest-neighbor interactions

$$\hat{H}_{\text{n.n.}} = -\frac{J}{2} \sum_{\langle ij \rangle} \left(\frac{\hat{S}_i^+ \hat{S}_j^- + \hat{S}_i^- \hat{S}_j^+}{2} + \hat{S}_i^z \hat{S}_j^z \right) . \qquad (4.5.1)$$

Inserting Eqs. 4.3.4 generally, we obtain

$$\hat{H}_{\text{n.n.}} = -\frac{J}{2} \sum_{\langle ij \rangle} \left[S\sqrt{1 - \frac{\hat{a}_i^{\dagger}\hat{a}_i}{2S}} \hat{a}_i \hat{a}_j^{\dagger} \sqrt{1 - \frac{\hat{a}_j^{\dagger}\hat{a}_j}{2S}} + S\hat{a}_i^{\dagger}\sqrt{1 - \frac{\hat{a}_i^{\dagger}\hat{a}_i}{2S}}\sqrt{1 - \frac{\hat{a}_j^{\dagger}\hat{a}_j}{2S}}\hat{a}_j \right]$$
$$- \frac{J}{2} \sum_{\langle ij \rangle} \left(S - \hat{a}_i^{\dagger}\hat{a}_i\right)\left(S - \hat{a}_j^{\dagger}\hat{a}_j\right) .$$

We now keep the first two terms in the expansion of $\phi(\hat{n}_i)$, see Eq. 4.4.1. Therefore, simply inserting into Eq. 4.5.1

$$\hat{H}_{n.n.} \approx -\frac{J}{2} \sum_{\langle ij \rangle} S \left(1 - \frac{\hat{a}_i^\dagger \hat{a}_i}{4S} \right) \hat{a}_i \hat{a}_j^\dagger \left(1 - \frac{\hat{a}_j^\dagger \hat{a}_j}{4S} \right) \tag{4.5.2}$$

$$-\frac{J}{2} \sum_{\langle ij \rangle} S \hat{a}_i^\dagger \left(1 - \frac{\hat{a}_i^\dagger \hat{a}_i}{4S} \right) \left(1 - \frac{\hat{a}_j^\dagger \hat{a}_j}{2S} \right) \hat{a}_j \tag{4.5.3}$$

$$-\frac{J}{2} \sum_{\langle ij \rangle} \left(S - \hat{a}_i^\dagger \hat{a}_i \right) \left(S - \hat{a}_j^\dagger \hat{a}_j \right) . \tag{4.5.4}$$

To be consistent with the approximation, we keep terms with up to four creation/annihilation operators in Eq. 4.5.2. One obtains

$$\hat{H}_{n.n.} \approx -Nz \frac{JS^2}{2} + JS \sum_{\langle ij \rangle} \left(\hat{a}_i^\dagger \hat{a}_i + \hat{a}_j^\dagger \hat{a}_j - \hat{a}_i^\dagger \hat{a}_j - \hat{a}_j^\dagger \hat{a}_i \right) \tag{4.5.5}$$

$$- J \sum_{\langle ij \rangle} \left[\hat{a}_i^\dagger \hat{a}_i \hat{a}_j^\dagger \hat{a}_j - \frac{1}{4} \left(\hat{a}_i^\dagger \hat{a}_i^\dagger \hat{a}_i \hat{a}_j + \hat{a}_i^\dagger \hat{a}_j^\dagger \hat{a}_j \hat{a}_j + \hat{a}_j^\dagger \hat{a}_i^\dagger \hat{a}_i \hat{a}_i + \hat{a}_j^\dagger \hat{a}_j^\dagger \hat{a}_j \hat{a}_i \right) \right] \tag{4.5.6}$$

with the following terms in the expansion being of order $1/S$ or higher.

We already saw that the quadratic terms in Eq. 4.5.5 can be diagonalized by going to the Fourier representation of the operators $\hat{a}_i$, after which one obtains the Hamiltonian 4.4.5—in this case with $B_0 = 0$ and $\omega(\mathbf{k})$ the corresponding one for nearest neighbors interaction. Here, we pay attention to the new terms, for simplicity we look at one of them, e.g., the term containing $\hat{a}_j^\dagger \hat{a}_i^\dagger \hat{a}_i \hat{a}_i$. We denote with Δ the nearest-neighbor vector. Hence,

$$\frac{J}{4} \sum_{\langle ij \rangle} \hat{a}_j^\dagger \hat{a}_i^\dagger \hat{a}_i \hat{a}_i = \frac{J}{4} \sum_{\mathbf{R}_j, \Delta} \hat{a}_j^\dagger \hat{a}_{j+\Delta}^\dagger \hat{a}_{j+\Delta} \hat{a}_{j+\Delta}$$

$$= \frac{J}{4N^2} \sum_{\mathbf{R}_j, \Delta} \sum_{\mathbf{k}_1, \mathbf{k}_2, \mathbf{k}_3, \mathbf{k}_4} \hat{a}_{\mathbf{k}_1}^\dagger e^{-i\mathbf{k}_1 \cdot \mathbf{R}_j} \hat{a}_{\mathbf{k}_2}^\dagger e^{-i\mathbf{k}_2 \cdot (\mathbf{R}_j + \Delta)} \times$$

$$\times \hat{a}_{\mathbf{k}_3} e^{i\mathbf{k}_3 \cdot (\mathbf{R}_j + \Delta)} \hat{a}_{\mathbf{k}_4} e^{i\mathbf{k}_4 \cdot (\mathbf{R}_j + \Delta)}$$

$$= \frac{J}{4N^2} \sum_{\mathbf{R}_j, \Delta} \sum_{\mathbf{k}_1, \mathbf{k}_2, \mathbf{k}_3, \mathbf{k}_4} e^{-i(\mathbf{k}_1 + \mathbf{k}_2 - \mathbf{k}_3 - \mathbf{k}_4) \cdot \mathbf{R}_j} \times$$

$$\times \hat{a}_{\mathbf{k}_1}^\dagger \hat{a}_{\mathbf{k}_2}^\dagger \hat{a}_{\mathbf{k}_3} \hat{a}_{\mathbf{k}_4} e^{-i(\mathbf{k}_2 - \mathbf{k}_3 - \mathbf{k}_4) \cdot \Delta}$$

$$= \frac{J}{4N} \sum_{\mathbf{k}_1, \mathbf{k}_2, \mathbf{k}_3, \mathbf{k}_4} \delta_{\mathbf{k}_1 + \mathbf{k}_2, \mathbf{k}_3 + \mathbf{k}_4} \hat{a}_{\mathbf{k}_1}^\dagger \hat{a}_{\mathbf{k}_2}^\dagger \hat{a}_{\mathbf{k}_3} \hat{a}_{\mathbf{k}_4} \times$$

$$\times \sum_\Delta e^{-i(\mathbf{k}_2 - \mathbf{k}_3 - \mathbf{k}_4) \cdot \Delta}$$

$$= \frac{J}{4N} \sum_{\mathbf{k}_1, \mathbf{k}_2, \mathbf{k}_3, \mathbf{k}_4} \delta_{\mathbf{k}_1 + \mathbf{k}_2, \mathbf{k}_3 + \mathbf{k}_4} \hat{a}_{\mathbf{k}_1}^\dagger \hat{a}_{\mathbf{k}_2}^\dagger \hat{a}_{\mathbf{k}_3} \hat{a}_{\mathbf{k}_4} \sum_\Delta e^{i\mathbf{k}_1 \cdot \Delta} .$$

The last sum is simply a function of $\mathbf{k}_1$, which can be given explicitly once the lattice is known. For example, for a cubic lattice of lattice constant a

$$\gamma(\mathbf{k}_1) = \sum_{\Delta} e^{i\mathbf{k}_1 \cdot \Delta} = e^{iak_{1x}} + e^{-iak_{1x}} + e^{iak_{1y}} + e^{-iak_{1y}} + e^{iak_{1z}} + e^{-iak_{1z}}$$

$$= 2\left[\cos\left(ak_{1x}\right) + \cos\left(ak_{1y}\right) + \cos\left(ak_{1y}\right)\right].$$

The term $\hat{a}_{\mathbf{k}_1}^\dagger \hat{a}_{\mathbf{k}_2}^\dagger \hat{a}_{\mathbf{k}_3} \hat{a}_{\mathbf{k}_4}$ corresponds to two magnons with momentum $\mathbf{k}_3$ and $\mathbf{k}_4$, respectively, being annihilated, and two magnons with momentum $\mathbf{k}_1$ and $\mathbf{k}_2$ being created in the interaction process. The Kronecker delta $\delta_{\mathbf{k}_1+\mathbf{k}_2,\mathbf{k}_3+\mathbf{k}_4}$ ensures conservation of momentum, $\mathbf{k}_1 + \mathbf{k}_2 = \mathbf{k}_3 + \mathbf{k}_4$. The other four-magnon interaction terms in 4.5.5 can be treated analogously. One can easily see that for long wavelength magnons (i.e., k small), the scattering cross section of such processes go as $(ka)^4$ and is therefore small. Further interaction terms (six-magnon, etc) are suppressed by factors of increasing order in $1/S$ [4].

Including dipole–dipole interactions has two main effects: (i) it modifies the dispersion relation $\omega(\mathbf{k})$, which in that case depends on the angle between the wavevector $\mathbf{k}$ and the equilibrium direction of the saturated spins, since the dipole–dipole interaction is anisotropic. This gives rise to a *spin-wave manifold*. (ii) New three-magnon momentum-conserving interaction terms, e.g., $\hat{a}_{\mathbf{k}_1}^\dagger \hat{a}_{\mathbf{k}_2}^\dagger \hat{a}_{\mathbf{k}_3}$ or $\hat{a}_{\mathbf{k}_1} \hat{a}_{\mathbf{k}_2} \hat{a}_{\mathbf{k}_3}^\dagger$ are allowed: one magnon can split into two, and vice versa.

Check Points

- Why do we get magnon–magnon interactions?
- What conservation rules do they fulfill and where do they come from?

Chapter 5
Magneto-Optical Effects

In this chapter we will explore the interaction between light and magnetism in magnetic insulators. The coupling mechanism is the Faraday effect, in which the plane of polarization of the light rotates as it goes through a magnetized material. In turn, the light exerts a very tiny effective "magnetic field" on the spins: this is called the *inverse Faraday effect* and it is an example of *backaction*. In what follows we will go back to the classical realm to obtain the coupling term. This will allows us to, by proper quantization of the classical coupling energy term, obtain a coupling Hamiltonian between magnons and the quanta of light, photons.

5.1 Electromagnetic Energy and Zero-Loss Condition

We go back now to the full Maxwell equations in matter, in contrast to the magnetostatic approximation we used throughout Chap. 1

$$\nabla \times \mathbf{H} = \frac{\partial \mathbf{D}}{\partial t} + \mathbf{j}_F \tag{5.1.1}$$

$$\nabla \times \mathbf{E} = -\frac{\partial \mathbf{B}}{\partial t} \tag{5.1.2}$$

$$\nabla \cdot \mathbf{D} = \rho \tag{5.1.3}$$

$$\nabla \cdot \mathbf{B} = 0. \tag{5.1.4}$$

These equations describe completely an electromagnetic system once we give the constitutive equations

$$D_i = \varepsilon_{ij} E_j \tag{5.1.5}$$

$$B_i = \mu_{ij} H_j$$

© The Author(s), under exclusive license to Springer Nature Switzerland AG 2019
S. Viola Kusminskiy, *Quantum Magnetism, Spin Waves, and Optical Cavities*,
SpringerBriefs in Physics, https://doi.org/10.1007/978-3-030-13345-0_5

where we used the Einstein convention of summation over repeated indices. These constitutive equations assume an instantaneous response of the system: the response does not depend on time and the system has no memory. This is referred to as a *dispersionless*. For an isotropic system, the permittivity and permeability tensors are diagonal and proportional to the identity, $\varepsilon_{ij} = \varepsilon_0 \varepsilon_r \delta_{ij}$, $\mu_{ij} = \mu_0 \mu_r \delta_{ij}$ and Eq. 5.1.5 reduce to the scalar versions, $\mathbf{D} = \varepsilon_0 \varepsilon_r \mathbf{E}$ and $\mathbf{B} = \mu_0 \mu_r \mathbf{H}$.

We will argue that we can represent the coupling of light and magnetization just by using the permittivity tensor. Our aim now is to obtain symmetry conditions on the permittivity tensor ε_{ij} in the presence of a static magnetization in the material where the light propagates. For that we will use conservation of electromagnetic energy, given by a continuity equation involving the energy flux density. The instantaneous electromagnetic power per unit area is given by the Poynting vector

$$\mathbf{P} = \mathbf{E} \times \mathbf{H}. \tag{5.1.6}$$

If we consider volume V bounded by a surface $\mathcal{S}$, the energy per unit time entering the volume is given by

$$-\oint_{\mathcal{S}} \mathbf{P} \cdot \mathrm{d}s$$

where ds is an area element with vector pointing outwards. This power can be stored in the volume in the form of an energy density W, or dissipated—we denominate the dissipated power P_d. We can hence write

$$-\int_V \nabla \cdot \mathbf{P}\mathrm{d}^3 r = \int_V \frac{\partial W}{\partial t} \mathrm{d}^3 r + \int_V P_d \mathrm{d}^3 r$$

where on the LHS we have used Stokes theorem. Since the volume is arbitrary, we have

$$\nabla \cdot \mathbf{P} + \frac{\partial W}{\partial t} + P_d = 0$$

which has the form of a continuity equation. It remains to identify the terms W and P_d in terms of the electromagnetic fields. For that we look at the Maxwell equations and we see that by taking the scalar product of 5.1.1 with $\mathbf{E}$, of 5.1.2 with $\mathbf{H}$, and subtracting 5.1.2 from 5.1.1, we can obtain an equation for the Poynting vector $\mathbf{P}$ by using the vector identity

$$\nabla \cdot (\mathbf{A} \times \mathbf{C}) = \mathbf{C} \cdot (\nabla \times \mathbf{A}) - \mathbf{A} \cdot (\nabla \times \mathbf{C}) . \tag{5.1.7}$$

Putting all together we obtain the *Poynting theorem*

$$\nabla \cdot (\mathbf{E} \times \mathbf{H}) + \mathbf{H} \cdot \frac{\partial \mathbf{B}}{\partial t} + \mathbf{E} \cdot \frac{\partial \mathbf{D}}{\partial t} + \mathbf{E} \cdot \mathbf{j}_F = 0 \tag{5.1.8}$$

$$\nabla \cdot (\mathbf{E} \times \mathbf{H}) + H_i \frac{\partial \left(\mu_{ij} H_j \right)}{\partial t} + E_i \frac{\partial \left(\varepsilon_{ij} D_j \right)}{\partial t} + \mathbf{E} \cdot \mathbf{j}_F = 0 , \tag{5.1.9}$$

where in the last line we have used Eq. 5.1.5. For dispersionless, isotropic media, we obtain

$$\nabla \cdot (\mathbf{E} \times \mathbf{H}) + \frac{\mu_0 \mu_r}{2} \frac{\partial H^2}{\partial t} + \frac{\varepsilon_0 \varepsilon_r}{2} \frac{\partial E^2}{\partial t} + \mathbf{E} \cdot \mathbf{j}_F = 0 \,,$$

from where we can identify

$$P_d = \mathbf{E} \cdot \mathbf{j}_F$$
$$W = \frac{\mu_0 \mu_r}{2} H^2 + \frac{\varepsilon_0 \varepsilon_r}{2} E^2$$

as the dissipated power density and instantaneous energy density stored in the magnetic and electric fields, respectively.

We are interested however in time-averaged quantities. To proceed further we consider for simplicity monochromatic fields in complex notation

$$\mathbf{E}(t) = \mathrm{Re}\left\{\mathbf{E}(\omega)e^{-i\omega t}\right\}$$
$$\mathbf{H}(t) = \mathrm{Re}\left\{\mathbf{H}(\omega)e^{-i\omega t}\right\} \,.$$

It can be easily shown that the time average of the product of two oscillating fields $A(t) = A_0 \cos(\omega t)$, $B(t) = B_0 \cos(\omega t + \phi)$ over one period $T = 2\pi/\omega$ is given simply by

$$\langle A(t)B(t)\rangle_T = \frac{1}{2}\mathrm{Re}\left\{\tilde{A}\tilde{B}^*\right\} \tag{5.1.10}$$

where

$$A(t) = \mathrm{Re}\left\{A_0 e^{-i\omega t}\right\} = \mathrm{Re}\left\{\tilde{A}e^{-i\omega t}\right\}$$
$$B(t) = \mathrm{Re}\left\{B_0 e^{-i\phi}e^{-i\omega t}\right\} = \mathrm{Re}\left\{\tilde{B}e^{-i\omega t}\right\} \,.$$

As a rule, one works with the complex fields and takes the real part at the end of the calculation. In an abuse of notation, the tilde notation is dropped. We will use Eq. 5.1.10 to obtain a time average of the Poynting theorem given in Eq. 5.1.8. For that we write Maxwell equations in frequency space, in particular

$$\nabla \times \mathbf{H} = -i\omega \mathbf{D} + \mathbf{j}_F \tag{5.1.11}$$
$$\nabla \times \mathbf{E} = i\omega \mathbf{B} \,. \tag{5.1.12}$$

To obtain the Poynting theorem in complex form, we take the complex conjugate of Eq. 5.1.11 and perform the scalar product with $\mathbf{E}$, and take the scalar product of Eq. 5.1.12 with $\mathbf{H}^*$. Subtracting the resulting equations and using again the vector identity 5.1.7 we obtain

$$\nabla \cdot \left(\mathbf{E} \times \mathbf{H}^*\right) + i\omega \left(\mathbf{E} \cdot \mathbf{D}^* - \mathbf{H}^* \cdot \mathbf{B}\right) + \mathbf{E} \cdot \mathbf{j}_F^* = 0 \,,$$

form where the time average is easily obtained as

$$\text{Re}\left\{\nabla \cdot (\mathbf{E} \times \mathbf{H}^*) + i\omega\left(\mathbf{E} \cdot \mathbf{D}^* - \mathbf{H}^* \cdot \mathbf{B}\right) + \mathbf{E} \cdot \mathbf{j}_F^*\right\} = 0. \tag{5.1.13}$$

For lossless media, $\langle \nabla \cdot (\mathbf{E} \times \mathbf{H}^*)\rangle_T$ must vanish, since all power that enters a volume must leave within one cycle. Moreover, if there are no free currents, the dissipated power $\mathbf{E} \cdot \mathbf{j}_F^*$ is also zero. Therefore, in lossless media

$$\text{Re}\left\{i\omega\left(\mathbf{E} \cdot \mathbf{D}^* - \mathbf{H}^* \cdot \mathbf{B}\right)\right\} = 0. \tag{5.1.14}$$

Moreover, in the frequency domain dispersive effects are included in a simple way by frequency-dependent permittivity and permeability tensors

$$\mathbf{D}(\omega) = \bar{\varepsilon}(\omega) \cdot \mathbf{E}(\omega)$$
$$\mathbf{B}(\omega) = \bar{\mu}(\omega) \cdot \mathbf{H}(\omega)$$

where the bar indicates that $\bar{\varepsilon}(\omega)$, $\bar{\mu}(\omega)$ are matrices. Equation 5.1.14 then can be written as [4, 12]

$$\frac{i\omega}{2}\left[\mathbf{E}^* \cdot \left(\bar{\varepsilon}^\dagger - \bar{\varepsilon}\right) \cdot \mathbf{E} + \mathbf{H}^* \cdot \left(\bar{\mu}^\dagger - \bar{\mu}\right) \cdot \mathbf{H}\right] = 0 \tag{5.1.15}$$

where † indicates complex conjugate and transpose: $(\varepsilon_{ij})^\dagger = \varepsilon_{ji}^*$ (note that the same expression can be obtained directly from Eq. 5.1.8 by replacing the real fields using $\text{Re}\{z\} = (z + z^*)/2$ and noting that $\langle zz\rangle_T = \langle z^*z^*\rangle_T = 0$). We deduce therefore that for lossless media

$$\bar{\varepsilon}^\dagger = \bar{\varepsilon} \tag{5.1.16}$$
$$\bar{\mu}^\dagger = \bar{\mu},$$

that is, the permittivity and permeability must be Hermitian matrices. Note that if the material is isotropic and $\bar{\varepsilon}(\omega)$, $\bar{\mu}(\omega)$ can be written as in principle complex scalars $\varepsilon(\omega) = \varepsilon'(\omega) + i\varepsilon''(\omega)$, $\mu(\omega) = \mu'(\omega) + i\mu''(\omega)$, the zero-loss condition implies that the imaginary parts $\varepsilon''(\omega)$ and $\mu''(\omega)$ must vanish.

To define the average electromagnetic energy density in the presence of dispersion is a little bit more subtle [12]. We give here for completeness the corresponding expression without proof

$$\langle W\rangle_T = \frac{1}{4}\left[\mathbf{E}^* \cdot \frac{\partial\left(\omega\bar{\varepsilon}\right)}{\partial\omega} \cdot \mathbf{E} + \mathbf{H}^* \cdot \frac{\partial\left(\omega\bar{\mu}\right)}{\partial\omega} \cdot \mathbf{H}\right]. \tag{5.1.17}$$

If $\bar{\varepsilon}$ and $\bar{\mu}$ are independent of frequency, this expression reduces to

$$\langle W\rangle_T = \frac{1}{4}\left[\mathbf{E}^* \cdot \bar{\varepsilon} \cdot \mathbf{E} + \mathbf{H}^* \cdot \bar{\mu} \cdot \mathbf{H}\right] \tag{5.1.18}$$

as expected.

1. *Exercise: Prove Eq.* 5.1.15 *starting from* 5.1.14.

Check Points

- Obtain Eq. 5.1.13 from Maxwell equations
- What is the zero loss condition?
- What does it tell us about the symmetries of the permittivity tensor?

5.2 Permittivity Tensor and Magnetization

In the following section we will use the permittivity tensor $\bar{\varepsilon}$ to describe the Faraday effect in a magnetized sample. For that we will use the symmetry properties of $\bar{\varepsilon}$ in the presence of a permanent magnetization, which breaks time reversal invariance. If we write the permittivity tensor explicitly separating the real and imaginary parts

$$\varepsilon_{ij} = \varepsilon'_{ij} + i\varepsilon''_{ij},$$

the zero-loss condition 5.1.16 tells us that real and imaginary parts are respectively symmetric and antisymmetric matrices:

$$\varepsilon'_{ij}(\mathbf{M}) = \varepsilon'_{ji}(\mathbf{M}) \tag{5.2.1}$$
$$\varepsilon''_{ij}(\mathbf{M}) = -\varepsilon''_{ji}(\mathbf{M}),$$

where we have made explicit a possible dependence on the magnetization $\mathbf{M}$. On the other hand, Onsager reciprocity relations for response functions dictate how the permittivity transforms under time reversal symmetry

$$\varepsilon'_{ij}(\mathbf{M}) = \varepsilon'_{ji}(-\mathbf{M}) \tag{5.2.2}$$
$$\varepsilon''_{ij}(\mathbf{M}) = \varepsilon''_{ji}(-\mathbf{M}),$$

where the time reversed form of $\bar{\varepsilon}$ consists in transposing the matrix and at the same time inverting the magnetization vector. We see therefore that the real and imaginary parts are also symmetric and antisymmetric in the magnetization. Putting all together we obtain

$$\varepsilon'_{ij}(\mathbf{M}) = \varepsilon'_{ji}(\mathbf{M}) = \varepsilon'_{ji}(-\mathbf{M}) \tag{5.2.3}$$
$$\varepsilon''_{ij}(\mathbf{M}) = -\varepsilon''_{ji}(\mathbf{M}) = \varepsilon''_{ji}(-\mathbf{M}).$$

In linear response, the permittivity depends linearly on the magnetization. This is valid as long as the effect of the magnetization on the permittivity is small. To fulfill conditions 5.2.3 to first order in the magnetization we write [12]

$$\varepsilon_{ij}(\mathbf{M}) = \varepsilon_0 \left(\varepsilon_r \delta_{ij} - i f \epsilon_{ijk} M_k\right) \tag{5.2.4}$$

where we have assumed the material is isotropic, and f is a small material-dependent parameter related to the Faraday rotation as we show below. A material for which the permittivity takes this form is denominated *gyrotropic*.

1. **Exercise: Prove that Eq.** 5.2.4 *fulfills* 5.2.3.

Check Points

- Explain how Eq. 5.2.4 is obtained

5.3 Faraday Effect

For optical frequencies one can usually safely take the permeability of a dielectric as the vacuum permeability μ_0, even for a magnetic material. This amounts to neglecting the interaction of the small oscillating magnetic field part of the optical electromagnetic field with the material. In turn, the interaction between the electric part of the optical field and the magnetization in the sample is modeled by the permittivity given in Eq. 5.2.4. The magnetization $\mathbf{M}$, even if it has a time dependence, it it much slower than the optical fields and therefore it is well defined.

To understand how the magnetization dependent permittivity in Eq. 5.2.4 encapsulates the Faraday effect we will first derive the Fresnel equation for the optical field, starting from the Maxwell equations 5.1.11 and 5.1.12 in the absence of free currents, $\mathbf{j}_F = 0$. We are interested in light propagating through a material, we therefore write the electric and magnetic fields as plane waves of frequency ω and wavevector $\mathbf{k}$

$$\mathbf{E}(t, \mathbf{r}) = \mathbf{E} e^{-i(\omega t - \mathbf{k} \cdot \mathbf{r})}$$
$$\mathbf{H}(t, \mathbf{r}) = \mathbf{H} e^{-i(\omega t - \mathbf{k} \cdot \mathbf{r})}.$$

In momentum and frequency representation, Eqs. 5.1.11 and 5.1.12 read

$$\mathbf{k} \times \mathbf{E} = \mu_0 \omega \mathbf{H} \tag{5.3.1}$$
$$\mathbf{k} \times \mathbf{H} = -\omega \mathbf{D}. \tag{5.3.2}$$

Inserting Eq. 5.3.1 into 5.3.2, using $\mathbf{D} = \bar{\varepsilon} \cdot \mathbf{E}$ and the product rule $\mathbf{a} \times (\mathbf{b} \times \mathbf{c}) = \mathbf{b}(\mathbf{a} \cdot \mathbf{c}) - \mathbf{c}(\mathbf{a} \cdot \mathbf{b})$ one obtains

$$\frac{k^2}{\mu_0 \omega^2} \left[\mathbf{E} - \frac{\mathbf{k}(\mathbf{k} \cdot \mathbf{E})}{k^2}\right] = \bar{\varepsilon} \cdot \mathbf{E}. \tag{5.3.3}$$

This is a form of the Fresnel equation, and it determines the dispersion relation of the electromagnetic wave (that is, $\omega(\mathbf{k})$) by imposing the determinant of its coefficients to be zero. In components

$$\frac{k^2}{\mu_0 \omega^2} \left[\delta_{ij} E_j - \frac{k_i k_j E_j}{k^2} \right] = \varepsilon_{ij} E_j$$

and therefore

$$\det \left\{ \frac{k^2}{\mu_0 \omega^2} \left[\delta_{ij} - \frac{k_i k_j}{k^2} \right] - \varepsilon_{ij} \right\} = 0 .$$

The term $\mathbf{k}(\mathbf{k} \cdot \mathbf{E})/k^2$ gives us simply the projection of the electric field along the propagation direction $\mathbf{k}/k$. Whereas in vacuum the electric field $\mathbf{E}$ is purely transverse, in a medium the purely transverse field is actually $\mathbf{D}$, see Eq. 5.3.2. We will work now however with the particular form of the permittivity given in Eq. 5.2.4, and consider the direction of propagation $\mathbf{k}/k$ to coincide with the magnetization axis, $\mathbf{M} = M\hat{z}$, in which case the $\mathbf{E}$ field is also transverse as one can easily verify by using the resulting form of the permittivity tensor

$$\bar{\varepsilon} = \varepsilon_0 \begin{pmatrix} \varepsilon_r & -ifM & 0 \\ ifM & \varepsilon_r & 0 \\ 0 & 0 & \varepsilon_r \end{pmatrix} . \tag{5.3.4}$$

The Fresnel equation reduces therefore to

$$\left| \begin{pmatrix} \frac{k^2}{\mu_0 \omega^2} - \varepsilon_0 \varepsilon_r & -ifM \\ ifM & \frac{k^2}{\mu_0 \omega^2} - \varepsilon_0 \varepsilon_r \end{pmatrix} \right| = 0$$

with solutions

$$k_\pm^2 = \left(\frac{\omega}{c} \right)^2 (\varepsilon_r \pm fM) \tag{5.3.5}$$

wehere we have used that $c = 1/\sqrt{\varepsilon_0 \mu_0}$. Inserting these solutions back into Eq. 5.3.3 with $\bar{\varepsilon}$ given by Eq. 5.3.4 (in this case only two dimensional, since $E_z = 0$) we obtain

$$E_x = \mp i E_y ,$$

with $\mp$ corresponding to $k_\pm^2$. We have therefore obtained two solutions for the propagating wave along $\hat{z}$, with the same amplitude and cicularly polarized in the xy plane, but with opposite polarizations for k_+ and k_-:

$$\mathbf{E}_\pm(z, t) = E_0 \operatorname{Re} \left\{ \left(\hat{\mathbf{e}}_x \pm i\hat{\mathbf{e}}_y \right) e^{i(k_\pm z - \omega t)} \right\} . \tag{5.3.6}$$

To derive the Faraday rotation we consider now an EM wave propagating in the medium such that at $z = 0$ it is linearly polarized along $\hat{x}$ with amplitude E_0,

$$\mathbf{E}(z = 0, t) = E_0 \hat{\mathbf{e}}_x e^{-i\omega t} . \tag{5.3.7}$$

The linear combination of Eq. 5.3.6

$$\mathbf{E}(z, t) = \frac{1}{2} \left[\mathbf{E}_+(z, t) + \mathbf{E}_-(z, t) \right]$$

fulfills the condition 5.3.7. We therefore have as solution for the propagating wave $\mathbf{E}(z, t) = \mathrm{Re} \left\{ E_x \hat{\mathbf{e}}_x + E_y \hat{\mathbf{e}}_y \right\}$

$$E_x = \frac{E_0}{2} \left(e^{ik_+ z} + e^{ik_- z} \right)$$

$$E_y = i \frac{E_0}{2} \left(e^{ik_+ z} - e^{ik_- z} \right) .$$

After the wave propagated a distance L through the material

$$\frac{E_y}{E_x} = \tan \left[\left(\frac{k_- - k_+}{2} \right) L \right] ,$$

which indicates that the plane of polarization of the light rotated by an amount $\theta_F L$, where

$$\theta_F = \frac{k_- - k_+}{2}$$

is the *Faraday rotation per unit length*. This is depicted schematically in Fig. 5.1. Using Eq. 5.3.5 and $f M \ll \varepsilon_r$ we obtain

$$\theta_F = \frac{\omega}{2c\sqrt{\epsilon_r}} f M . \tag{5.3.8}$$

Check Points

- What is the Faraday effect?

Fig. 5.1 Vertically polarized light rotates its angle of polarization as it goes through a magnetized material

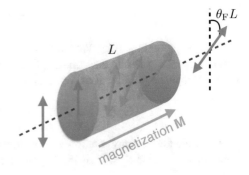

5.4 Magneto-Optical Energy

In the previous sections we saw that the magnetization in a medium modifies the permittivity tensor, which acquires an antisymmetric imaginary term due to the breaking of time reversal symmetry. As light goes through the material, it experiences this effective permittivity which we saw leads to the Faraday effect. Light and magnetization in the sample are therefore coupled. To obtain this coupling, we look at the electromagnetic energy obtained from Eq. 5.1.18 by using the permittivity Eq. 5.2.4. We see that the magnetization-dependent part of the permittivity introduces a correction to the usual electromagnetic energy expression, given by

$$U_{\mathrm{MO}} = -\frac{i}{4} f \varepsilon_0 \int \mathrm{d}^3 r \mathbf{M}(\mathbf{r}) \cdot \left[\mathbf{E}^*(\mathbf{r}) \times \mathbf{E}(\mathbf{r}) \right] . \tag{5.4.1}$$

One can easily prove that this term is real. In terms of the Faraday rotation per unit length this can be rewritten as

$$U_{\mathrm{MO}} = \theta_{\mathrm{F}} \sqrt{\frac{\varepsilon_r}{\varepsilon_0}} \int \mathrm{d}^3 r \frac{\mathbf{M}(\mathbf{r})}{M} \cdot \frac{\varepsilon_0}{2i\omega} \left[\mathbf{E}^*(\mathbf{r}) \times \mathbf{E}(\mathbf{r}) \right] . \tag{5.4.2}$$

The term

$$\mathbf{S}_{\mathrm{light}}(\mathbf{r}) = \frac{\varepsilon_0}{2i\omega} \left[\mathbf{E}^*(\mathbf{r}) \times \mathbf{E}(\mathbf{r}) \right] \tag{5.4.3}$$

is called the *optical spin density and* it is related to the *helicity* of light. For example, for circularly polarized light $\mathbf{S}_{\mathrm{light}}$ points perpendicular to the plane of polarization with a direction given by the right-hand-rule.

We know from the previous section that the magnetization causes the plane of polarization of the light to rotate. From Eq. 5.4.1 we see that the light itself acts as an effective magnetic field on the magnetization (compare with the usual expression for the Zeeman energy). This gives rise to the *inverse Faraday effect*, which takes into account the effect of the light on the magnetization dynamics of the sample. This effect is usually small, one can show that the effective light-induced magnetic field for YIG (Yttrium Iron Garnet, a magnetic insulator, widely used in both technical applications and current experiments) is of the order of 10^{-11}T per photon/μ^3. For comparison, the earth's magnetic field is of the order of 10^{-6}T! We will see however in the following chapter that this value can be enhanced by using an optical cavity, effectively "trapping" photons.

1. **Exercise: Derive Eq.** 5.4.2 **starting from** 5.1.18.

Check Points

- Obtain the correction to the electromagnetic energy if a medium is magnetized.

Chapter 6
Modern Topics: Cavity Optomagnonics

In this last chapter we will put together all concepts we learned so far to derive the basics of a topic of current research: Cavity Optomagnonics. In these systems, a magnetic insulator material forms a cavity for the light, which is used to enhance the magnon–photon coupling. We call this the *optomagnonic coupling*. As we pointed out in the previous chapter, the Faraday effect is usually a small effect, which depends on the Faraday rotation constant of the material and also on the path's length of the light inside of the material. Since in a cavity the light is "trapped", this effectively enhances the path's length and therefore the coupling. This is an intuitive way of seeing the enhancement, we will see this more formally in the next sections. For that we need first to learn about optical cavities and the quantization of the electromagnetic field.

6.1 Quantization of the Electromagnetic Field

We start by quantizing a single-mode field in a cavity formed by perfectly conducting walls at $z = 0$ and $z = L$. We summarize here the most important concepts, a more thorough discussion can be found for example in Refs. [13, 14]. We assume further the electric field to be polarized along x, $\mathbf{E} = E_x(z, t)\hat{\mathbf{e}}_x$. The boundary condition therefore implies

$$E_x(z = 0, t) = E_x(z = L, t) = 0. \tag{6.1}$$

From Maxwell equations in vacuum and no sources,

$$\nabla \times \mathbf{E} = -\frac{\partial \mathbf{B}}{\partial t} \qquad\qquad \nabla \cdot \mathbf{B} = 0 \tag{6.2}$$

$$\nabla \times \mathbf{B} = \mu_0 \varepsilon_0 \frac{\partial \mathbf{E}}{\partial t} \qquad\qquad \nabla \cdot \mathbf{E} = 0 \tag{6.3}$$

© The Author(s), under exclusive license to Springer Nature Switzerland AG 2019
S. Viola Kusminskiy, *Quantum Magnetism, Spin Waves, and Optical Cavities*,
SpringerBriefs in Physics, https://doi.org/10.1007/978-3-030-13345-0_6

we obtain a wave equation for the electric field

$$\nabla^2 \mathbf{E} = \mu_0 \varepsilon_0 \frac{\partial^2 \mathbf{E}}{\partial t^2} \tag{6.4}$$

which in terms of $\mathbf{E} = E_x(z, t)\hat{\mathbf{e}}_x$ simplifies to

$$\frac{\partial^2 E_x(z, t)}{\partial z^2} - \frac{1}{c^2} \frac{\partial^2 E_x(z, t)}{\partial t^2} = 0 . \tag{6.5}$$

The solution of Eq. 6.5 satisfying the boundary conditions in Eq. 6.1 is simply

$$E_x(z, t) = \sqrt{\left(\frac{2\omega_n^2}{V\varepsilon_0}\right)} q(t) \sin(k_n z) \tag{6.6}$$

with

$$\frac{\omega_n}{c} = k_n = \frac{\pi n}{L} . \tag{6.7}$$

In Eq. 6.6, $q(t)$ has units of length and $V = LS$ is the volume of the cavity, where S is its cross-section. From the left Eq. in 6.3 one obtains $\mathbf{B} = B_y(z, t)\hat{\mathbf{e}}_y$ with

$$B_y(z, t) = \frac{\mu_0 \varepsilon_0}{k_n} \sqrt{\left(\frac{2\omega_n^2}{V\varepsilon_0}\right)} \dot{q}(t) \cos(k_n z) . \tag{6.8}$$

Inserting Eqs. 6.6 and 6.8 into the electromagnetic energy

$$E_{\text{EM}} = \frac{1}{2} \int dV \left(\varepsilon_0 \mathbf{E}^2 + \frac{1}{\mu_0} \mathbf{B}^2\right)$$

one obtains

$$E_{\text{EM}} = \frac{1}{2} \left(\omega_n^2 q^2 + p^2\right) \tag{6.9}$$

where we have defined $p = \dot{q}$ the canonical momentum of a "particle" of unit mass. Eq. 6.9 is the energy of a harmonic oscillator of unit mass. We can now proceed to quantize the theory by

$$q \to \hat{q}$$
$$p \to \hat{p}$$

and imposing the commutator $[\hat{q}, \hat{p}] = i\hbar$. It is convenient to introduce the bosonic creation and annihilation operators

$$\hat{a} = \frac{1}{\sqrt{2\hbar\omega_n}} \left(\omega_n \hat{q} + i\hat{p}\right)$$

$$\hat{a}^\dagger = \frac{1}{\sqrt{2\hbar\omega_n}} \left(\omega_n \hat{q} - i\hat{p}\right)$$

which satisfy $[\hat{a}, \hat{a}^\dagger]$ in terms of which the electric and magnetic field can be expressed as

$$\hat{E}_x(z, t) = \sqrt{\left(\frac{\hbar\omega_n}{V\varepsilon_0}\right)} \left(\hat{a} + \hat{a}^\dagger\right) \sin(k_n z) \tag{6.10}$$

$$B_y(z, t) = \sqrt{\left(\frac{\hbar\omega_n}{V\varepsilon_0}\right)} \left(\hat{a} - \hat{a}^\dagger\right) \cos(k_n z). \tag{6.11}$$

The energy Eq. 6.9 gives rise to the usual harmonic-oscillator Hamiltonian

$$\hat{H}_{EM} = \hbar\omega_n \left(\hat{a}^\dagger\hat{a} + \frac{1}{2}\right).$$

The time dependence of the ladder operators can be obtained from the Heisenberg equation of motion and is given by

$$\hat{a}(t) = \hat{a}(0)e^{-i\omega t}$$
$$\hat{a}^\dagger(t) = \hat{a}^\dagger(0)e^{i\omega t}.$$

Note that an eigenstate of the number operator $\hat{n} = \hat{a}^\dagger\hat{a}$, $\hat{n}|n\rangle = n|n\rangle$, is an energy eigenstate but the electric field operator's expectation value vanishes

$$\langle n|\hat{E}_x|n\rangle = 0$$

and therefore it is not well defined. The expectation value squared field $\langle n|\hat{E}_x^2|n\rangle$ is however finite, as one can easily prove. This reflects the uncertainty in the phase of the electric field, which is conjugate to the number operator. We call the excitation with energy $\hbar\omega_n$ a *photon*.

We now proceed to the quantization of multimode fields in a 3D cavity. For that, it is convenient to use the Coulomb gauge

$$\nabla \cdot \mathbf{A}(\mathbf{r}, t) = 0, \tag{6.12}$$

in which both electric and magnetic field can be expressed in terms of the vector potential $\mathbf{A}(\mathbf{r}, t)$

$$\mathbf{E}(r, t) = -\frac{\partial \mathbf{A}(\mathbf{r}, t)}{\partial t} \tag{6.13}$$

$$\mathbf{B}(r, t) = \nabla \times \mathbf{A}(\mathbf{r}, t). \tag{6.14}$$

From Maxwell's equations, one obtains a wave equation for the vector potential

$$\nabla^2 \mathbf{A} - \frac{1}{c^2}\frac{\partial^2 \mathbf{A}}{\partial t} = 0$$

which can be solved by separating time end space variables. It is customary to split the time dependence

$$\mathbf{A}(\mathbf{r}, t) = \mathbf{A}^+(\mathbf{r}, t) + \mathbf{A}^-(\mathbf{r}, t)$$

such that

$$\mathbf{A}^+(\mathbf{r}, t) = \sum_k c_k \mathbf{u}_k(\mathbf{r}) e^{-i\omega_k t}$$

$$\mathbf{A}^-(\mathbf{r}, t) = \sum_k c_k^* \mathbf{u}_k^*(\mathbf{r}) e^{i\omega_k t}$$

where $\omega_k \geq 0$. The mode-functions $\mathbf{u}_k$ are solutions of

$$\left(\nabla^2 + \frac{\omega_k^2}{c^2}\right) \mathbf{u}_k(\mathbf{r}) = 0$$

satisfying orthonormality

$$\int \mathrm{d}V \mathbf{u}_k^*(\mathbf{r}) \mathbf{u}_{k'}(\mathbf{r}) = \delta_{k,k'}$$

and the appropriate boundary conditions. The index k indicates both the mode and the polarization vector. Moreover, due to the Coulomb gauge

$$\nabla \cdot \mathbf{u}_k(\mathbf{r}) = 0. \tag{6.15}$$

Guided by our one-mode example, we quantize replacing the amplitudes in the sum over modes by creation and annihilation operators

$$\hat{\mathbf{A}}(\mathbf{r}, t) = \sum_k \sqrt{\frac{\hbar}{2\omega_k \varepsilon_0}} \left[\hat{a}_k \mathbf{u}_k(\mathbf{r}) e^{-i\omega_k t} + \hat{a}_k^\dagger \mathbf{u}_k\hat{\ }^*(\mathbf{r}) e^{i\omega_k t}\right]$$

from which, using Eq. 6.13, we obtain the electric field operator

$$\hat{\mathbf{E}}(\mathbf{r}, t) = \hat{\mathbf{E}}^+(\mathbf{r}, t) + \hat{\mathbf{E}}^-(\mathbf{r}, t) = i \sum_k \sqrt{\frac{\hbar\omega_k}{2\varepsilon_0}} \left[\hat{a}_k \mathbf{u}_k(\mathbf{r}) e^{-i\omega_k t} - \hat{a}_k^\dagger \mathbf{u}_k\hat{\ }^*(\mathbf{r}) e^{i\omega_k t}\right]$$

$$\tag{6.16}$$

with the same convention for $\hat{\mathbf{E}}^\pm(\mathbf{r}, t)$ as for $\hat{\mathbf{A}}^\pm(\mathbf{r}, t)$. Using Eq. 6.14 we can obtain the corresponding expression for the magnetic field. The bosonic ladder operators $\hat{a}_k, \hat{a}_k^\dagger$ as defined are dimensionless and satisfy the usual commutation relations

$$\left[\hat{a}_k, \hat{a}_{k'}\right] = \left[\hat{a}_k^\dagger, \hat{a}_{k'}^\dagger\right] = 0$$

$$\left[\hat{a}_k, \hat{a}_{k'}^\dagger\right] = \delta_{k,k'}$$

and the corresponding Hamiltonian is that of a collection of non-interacting harmonic oscillators

$$H_{EM} = \sum_k \hbar\omega_k \left(\hat{a}_k^\dagger \hat{a}_k + \frac{1}{2} \right) .$$

The sum over modes has no cutoff and therefore the factor $1/2$ leads to a divergence. In these notes we will not worry about that, since we will always work with differences of energies, where this factor cancels out. We will therefore simply omit this term in the following. The energy eigenstates are also eigenstates of the number operator $\hat{n}_k = \hat{a}_k^\dagger \hat{a}_k$

$$\hat{n}_k |n_k\rangle = n_k |n_k\rangle$$

$$E_{n_k} = \hbar\omega_k \left(n_k + \frac{1}{2} \right) .$$

The states $|n_k\rangle$ form an orthonormal basis of the Hilbert space and are called number or *Fock* states, where n_k gives the number of photons in state k. A general multi-mode state can be written as

$$|\psi\rangle = \sum_{n_1, n_2, \ldots} c_{n_1, n_2, \ldots} |n_1, n_2, \ldots\rangle .$$

In the simplest example, one uses periodic boundary conditions in a cubic box of side length L. In this case,

$$\mathbf{u}_k(\mathbf{r}) = \frac{1}{L^{3/2}} \hat{\mathbf{e}}^\lambda e^{i\mathbf{k}\cdot\mathbf{r}}$$

with $\mathbf{k} = 2\pi/L \left(n_x, n_y, n_z \right)$, $n_i = 0, \pm1, \pm2, \ldots$ and $\lambda = 1, 2$ indicating the polarization, which from Eq. 6.15 must fulfill

$$\hat{\mathbf{e}}^\lambda \cdot \mathbf{k} = 0 .$$

One can analogously use reflecting boundary conditions, where the solutions are standing waves as in the single mode we studied above. Note that in this case the normalization factor of $\hat{\mathbf{E}}(\mathbf{r}, t)$ and the quantization of $\mathbf{k}$ will be different.

1. *Exercise: Derive Eq. 6.4*
2. *Exercise: Derive Eq. 6.9*

Check Points

- Write a general expression for the quantized electric field.

6.2 Optical Cavities as Open Quantum Systems

Cavities are usually open systems, in contact with an environment both because, for example, the mirrors are not perfect, allowing contact to a bath of photons or/and phonons, and because we want to have access to the cavity by means of an external probe. The environment (also called bath, or reservoir) is assumed to be very large and in thermal equilibrium, and it is modeled as a collection of harmonic oscillators. The simplest Hamiltonian of the cavity plus bath is written as

$$\hat{H} = \hat{H}_S + \hat{H}_R + \hat{H}_I$$

where $\hat{H}_S$, $\hat{H}_R$, and $\hat{H}_I$ are the cavity, reservoir, and interaction Hamiltonians respectively

$$\hat{H}_S = \hbar\Omega\hat{a}^\dagger\hat{a}$$
$$\hat{H}_R = \sum_k \hbar\omega_k \hat{b}_k^\dagger \hat{b}_k$$
$$\hat{H}_I = \hbar \sum_k \left(g_k \hat{a}^\dagger \hat{b}_k + g_k^* \hat{b}_k^\dagger \hat{a} \right) ,$$

where we have taken for simplicity only one mode for the cavity $\hat{a}$ with frequency Ω. The interaction Hamiltonian represents an excitation in the cavity being converted into one in the reservoir and vice-versa, with coupling constant g_k.

Our aim is to obtain an effective equation of motion for the cavity mode $\hat{a}$ which encapsulates the effect of the bath [15]. This procedure is denominated to *integrate out* the bath, and it means that, since we are not interested in the dynamics of the bath per se, we want to eliminate these degrees of freedom and just retain the ones we are interested in, in this case, the single cavity mode. We begin by writing the Heisenberg equations of motion for both the cavity mode and the reservoir modes

$$\dot{\hat{a}}(t) = -i\Omega\hat{a}(t) - i\sum_k g_k \hat{b}_k(t) \tag{6.17}$$

$$\dot{\hat{b}}_k(t) = -i\omega_k \hat{b}(t) - i g_k^* \hat{a}(t) . \tag{6.18}$$

We can integrate formally Eq. 6.18 to obtain

$$\hat{b}_k(t) = \hat{b}_k(0)e^{-i\omega_k t} - i g_k^* \int_0^t dt' \hat{a}(t')e^{-i\omega_k(t-t')} , \tag{6.19}$$

where the first term corresponds to the free evolution of $\hat{b}_k$ and the second one is due to the interaction with the cavity. Substituting Eq. 6.19 into 6.17 we obtain

$$\dot{\hat{a}}(t) = -i\Omega\hat{a}(t) - i\sum_k g_k \hat{b}_k(0)e^{-i\omega_k t} - \sum_k |g_k|^2 \int_0^t dt' \hat{a}(t')e^{-i\omega_k(t-t')} . \tag{6.20}$$

We now transform the cavity operators to a rotating frame with frequency Ω

$$\hat{A}(t) = \hat{a}(t)e^{i\Omega t} .$$

We see that this transformation preserves the bosonic commutation relations

$$\left[\hat{A}(t), \hat{A}(t)\right] = \left[\hat{A}^\dagger(t), \hat{A}^\dagger(t)\right] = 0$$
$$\left[\hat{A}(t), \hat{A}^\dagger(t)\right] = 1$$

and removes the free, fast rotating term from the equation of motion:

$$\dot{\hat{A}}(t) = -i\sum_k g_k \hat{b}_k(0)e^{-i(\Omega-\omega_k)t} - \sum_k |g_k|^2 \int_0^t dt' \hat{A}(t')e^{-i(\Omega-\omega_k)(t-t')} . \quad (6.21)$$

The first term in Eq. 6.21 is denominated the *noise operator*

$$\hat{F}(t) = -i\sum_k g_k \hat{b}_k(0)e^{-i(\Omega-\omega_k)t} .$$

We see that this operator is composed of many different frequencies and therefore oscillates rapidly in time. Its effect on the cavity mode is that of exerting random "quantum kicks". Its expectation value for a reservoir in thermal equilibrium is easily shown to be zero

$$\langle \hat{F}(t)\rangle_R = 0$$

and therefore this operator is the quantum analog to the noise due to the environment responsible for the Brownian motion of a classical particle. The second term in Eq. 6.21

$$\hat{B}_{ba} = -\sum_k |g_k|^2 \int_0^t dt' \hat{A}(t')e^{-i(\Omega-\omega_k)(t-t')} \quad (6.22)$$

is due to *backaction*: changes in the cavity mode affect slightly the bath, which in turn acts back onto the cavity. We will see in the following that this term leads to decay of the cavity mode, which corresponds to dissipation of energy from the cavity into the environment.

To analyze the second term in Eq. 6.21 we use that the environment volume is large, which allows us to take the continuum limit for the bath modes. We write the sum over modes directly in terms of a *density of states* (DOS) which we do not specify, this will depend on the details of the bath. The DOS $\mathcal{D}(\omega_k)$ gives the number of modes with frequency between ω_k and $\omega_k + d!_k$. In terms of the DOS, the second term in Eq. 6.21 is

$$\hat{B}_{ba} = -\int_0^\infty d\omega_k \mathcal{D}(\omega_k) |g(\omega_k)|^2 \int_0^t dt' \hat{A}(t')e^{-i(\Omega-\omega_k)(t-t')} . \quad (6.23)$$

To perform this integral we have to resort to approximations that rely on the physics of the system. We will work with what is known as the Weisskopf-Wigner approximation, which is at its core a Markovian approximation: the evolution of the system of interest is local in time. This implies a separation of time scales between the bath, which we assume to be the fast, and the system (in our case, the cavity mode) which is slow. The information from the system that goes into the reservoir is lost, since the bath fluctuates very rapidly. The system therefore is said to have *no memory*. The timescale of the bath is defined by the inverse bandwidth $1/W$. If the rate of variation $\hat{A}(t)$ is slow compared to this timescale, we can replace $\hat{A}(t') \rightarrow \hat{A}(t)$ in Eq. 6.23, and extend the integral from t to ∞

$$\hat{B}_{ba} \approx -\int_0^\infty d\omega_k \mathcal{D}(\omega_k) |g(\omega_k)|^2 \hat{A}(t) \int_0^\infty d\tau e^{-i(\Omega - \omega_k)\tau}$$

where we defined $\tau = t - t'$. We use now that

$$\int_0^\infty d\tau e^{-i(\Omega - \omega_k)\tau} = \pi\delta(\omega_k - \Omega) - i\mathcal{P}\left(\frac{1}{\omega_k - \Omega}\right),$$

where the last term indicates the principal part. We neglect this term for the moment, since it leads to a frequency shift (note that its contribution is proportional to the cavity operator, and has an i in front). Evaluating the Delta function we obtain

$$\hat{B}_{ba} \approx -\hat{A}(t)\pi\mathcal{D}(\Omega)|g(\Omega)|^2$$

and therefore the effective equation of motion for the cavity mode in the rotating frame is

$$\dot{\hat{A}}(t) = -\frac{\gamma}{2}\hat{A}(t) + \hat{F}(t), \tag{6.24}$$

with

$$\gamma = \pi\mathcal{D}(\Omega)|g(\Omega)|^2$$

the *cavity decay rate*. Eq. 6.24 is a *quantum Langevin equation* and the decay rate γ and the noise operator $\hat{F}(t)$ can be shown to fulfill the *fluctuation-dissipation theorem*

$$\gamma = \frac{1}{\bar{n}} \int_{-\infty}^\infty d\tau \langle \hat{F}^\dagger(\tau)\hat{F}(0)\rangle_R$$

in equilibrium, with

$$\bar{n} = \langle \hat{b}^\dagger(\Omega)\hat{b}(\Omega)\rangle_R = \frac{1}{e^{\hbar\beta\Omega} - 1}$$

the thermal occupation of the bath at the cavity frequency. In the Markov approximation the noise correlators fulfill

$$\langle \hat{F}^\dagger(t')\hat{F}(t'')\rangle_R = \gamma\bar{n}\delta\left(t'-t''\right)$$
$$\langle \hat{F}(t')\hat{F}^\dagger(t'')\rangle_R = \gamma\left(\bar{n}+1\right)\delta\left(t'-t''\right).$$

In particular for the vacuum one obtains

$$\langle 0|\hat{F}(t')\hat{F}^\dagger(t'')|0\rangle_R = \gamma\delta\left(t'-t''\right).$$

This delta-correlated noise shows clearly that the dynamics of the bath is fast compared to that of the system of interest, and that it has no memory since every "quantum kick" is uncorrelated with the previous one. Usually the noise operator is normalized to an operator $\hat{A}_{in}$ such that

$$\langle 0|\hat{A}_{in}(t')\hat{A}_{in}^\dagger(t'')|0\rangle_R = \delta\left(t'-t''\right)$$

and the equation of motion is written as

$$\dot{\hat{A}}(t) = -\frac{\gamma}{2}\hat{A}(t) + \sqrt{\gamma}\hat{A}_{in}(t)$$

or, in the original frame,

$$\dot{\hat{a}}(t) = -i\Omega\hat{a}(t) - \frac{\gamma}{2}\hat{a}(t) + \sqrt{\gamma}\hat{a}_{in}(t).$$

For the decay rate sometimes $\kappa = \gamma/2$ is used.

Check Points

- Write the total Hamiltonian of a cavity coupled to an environment, and explain each term
- How does one obtain an effective equation of motion for the cavity mode? What approximations are involved?
- Write the effective equation of motion for the cavity mode
- What is the meaning of κ (or γ)?

6.3 The Optomagnonic Hamiltonian

We will now put together all the elements from the previous sections to derive the optomagnonic Hamiltonian, that is, the Hamiltonian for a system in which optical photons couple to magnons. For that, we will quantize the interaction term given by the Faraday effect, Eq. 5.4.1. This section and the next follow the recent work in Ref. [16].

We can quantize the electric field following Sect. 6.1, $\hat{\mathbf{E}}^+(\mathbf{r}, t) = \sum_\beta \mathbf{E}_\beta(\mathbf{r})\hat{a}_\beta(t)$ and correspondingly $\hat{\mathbf{E}}^-(\mathbf{r}, t) = \sum_\beta \mathbf{E}_\beta^*(\mathbf{r})\hat{a}_\beta^\dagger(t)$, where $\mathbf{E}_\beta(\mathbf{r})$ indicates the βth

eigenmode of the electric field (eigenmodes are indicated with greek letters in what follows). The magnetization requires more careful consideration, since $\mathbf{M}(\mathbf{r})$ depends on the local spin operator which, in general, cannot be written as a linear combination of bosonic modes. There are however two simple cases: (i) the spin-wave approximation, which is valid for small deviations of the spins from equilibrium and, as we saw in Sect. 4.4, the Holstein-Primakoff representation can be truncated to linear order in the bosonic magnon operators, and (ii) considering the homogeneous Kittel mode[1] $\mathbf{M}(\mathbf{r}) = \mathbf{M}$, for which we can work simply with the resulting macrospin $\mathbf{S}$. In the following we treat this second case. Although it is valid only for the homogeneous magnon mode, it allows us to capture the nonlinear dynamics of the spin.

From Eq. 5.4.1 we obtain the coupling Hamiltonian

$$\hat{H}_{MO} = \hbar \sum_{j\beta\gamma} \hat{S}_j G^j_{\beta\gamma} \hat{a}^\dagger_\beta \hat{a}_\gamma \qquad (6.25)$$

with coupling constants

$$G^j_{\beta\gamma} = -i \frac{\varepsilon_0 f M_s}{4\hbar S} \epsilon_{jmn} \int d\mathbf{r}\, E^*_{\beta m}(\mathbf{r}) E_{\gamma n}(\mathbf{r}), \qquad (6.26)$$

where we replaced $M_j/M_s = \hat{S}_j/S$, with S the extensive total spin (scaling like the magnetic mode volume). G^j are hermitian matrices which in general cannot be simultaneously diagonalized. For simplicity, in the following we treat the case of a strictly diagonal coupling to some optical eigenmodes ($G^j_{\beta\beta} \neq 0$ but $G^j_{\alpha\beta} = 0$).

As an example, we consider circular polarization (R/L) in the $y - z$-plane. In this case, the optical spin density is perpendicular to this plane, and therefore G^x is diagonal while $G^y = G^z = 0$. The Hamiltonian H_{MO} is then diagonal in the the basis of circularly polarized waves, $\mathbf{e}_{R/L} = \frac{1}{\sqrt{2}} (\mathbf{e}_y \mp i\mathbf{e}_z)$. We choose moreover the magnetization axis along the $\hat{\mathbf{z}}$ axis. This setup is shown schematically in Fig. 6.1. The rationale behind choosing the coupling direction *perpendicular* to the magnetization axis, is to maximize the coupling to the magnon mode, that is to the *deviations* of the magnetization with respect to the magnetization axis. The light field hence only couples to the x component of the spin operator, $\hat{S}_x$.

Again for simplicity, we consider the case of plane waves for quantizing the electric field. Therefore

$$\hat{\mathbf{E}}^{+(-)}(\mathbf{r}, t) = +(-)i \sum_j \mathbf{e}_j \sqrt{\frac{\hbar\omega_j}{2\varepsilon_0\varepsilon V}} \hat{a}^{(\dagger)}_j(t) e^{+(-)i\mathbf{k_j}\cdot\mathbf{r}},$$

where V is the volume of the cavity, $\mathbf{k}_j$ the wave vector of mode j and we have identified the positive and negative frequency components of the field as $\mathbf{E} \to \hat{\mathbf{E}}^+$,

---

[1]The Kittel mode is a spin wave with $\mathbf{k} = 0$, so that all spins precess in phase and can be replaced by a precessing macrospin.

Fig. 6.1 The Kittel mode
with frequency Ω is excited
on top of a homogeneous
magnetic ground state with
magnetization along the $\mathbf{e}_z$
axis. The same material
serves as the optical cavity.
The optical mode is right
circularly polarized in the
$y - z$ plane and the optical
spin density points along the
$\mathbf{e}_z$ axis

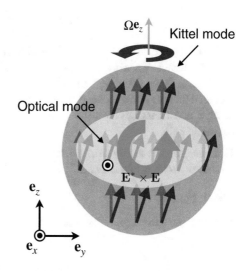

$\mathbf{E}^* \to \hat{\mathbf{E}}^-$. In the normalization of the fields we have used the relative permittivity ε of
the magnetic insulator, since the electric fields considered live in the cavity formed by
the material. The factor of $\varepsilon_0\varepsilon$ in the denominator ensures the normalization $\hbar\omega_j =
\varepsilon_0\varepsilon\langle j| \int d^3\mathbf{r}|\mathbf{E}(\mathbf{r})|^2|j\rangle - \varepsilon_0\varepsilon\langle 0| \int d^3\mathbf{r}|\mathbf{E}(\mathbf{r})|^2|0\rangle$, which corresponds to the energy
of a photon in state $|j\rangle$ above the vacuum $|0\rangle$. For two degenerate (R/L) modes at
frequency ω, using Eq. 5.3.8 we see that the frequency dependence cancels out and
we obtain the simple form for the optomagnonic Hamiltonian

$$H_{MO} = \hbar G \hat{S}_x (\hat{a}_L^\dagger \hat{a}_L - \hat{a}_R^\dagger \hat{a}_R)$$

with

$$G = \frac{1}{S} \frac{c}{4} \frac{\theta_F}{\sqrt{\varepsilon}}.$$

In general however an overlap factor $\xi \leq 1$ appears, which takes into account that
there is a mismatch between the optical and magnonic mode volumes. For example, current experiments couple optical whispering gallery modes (WGM) in a YIG
sphere to the magnetic Kittel mode [17–19]. The Kittel mode is a bulk mode, and
lives on the whole sphere, that is, the magnetic mode volume equals the volume
of the sphere. The WGMs live however very close to the surface, and therefore its
volume is smaller than the magnetic one, leading to and overlap factor $\xi < 1$,

$$G_{LL}^x = -G_{RR}^x = G = \frac{1}{S} \frac{c}{4} \frac{\theta_F}{\sqrt{\varepsilon}} \xi . \tag{6.27}$$

We now consider an incoming laser, which drives only one of the two circular
polarizations in the cavity. The total Hamiltonian of the cavity optomagnonic system
is therefore given simply by

$$H = -\hbar\Delta\hat{a}^\dagger\hat{a} - \hbar\Omega\hat{S}_z + \hbar G\hat{S}_x\hat{a}^\dagger\hat{a}\,, \tag{6.28}$$

where $\hat{a}^\dagger$ ($\hat{a}$) is the creation (annihilation) operator for the cavity mode photon which is being driven. We work in a frame rotating at the laser frequency ω_{las}, and $\Delta = \omega_{las} - \omega_{cav}$ is the detuning versus the optical cavity frequency ω_{cav}. In Eq. 6.28 we included that the dimensionless macrospin $\mathbf{S} = (S_x, S_y, S_z)$ has a magnetization axis along $\hat{\mathbf{z}}$, and a Larmor precession frequency Ω which can be controlled by an external magnetic field.

1. **Exercise: show that in the rotating frame the free Hamiltonian for a cavity driven mode is given by** $\hbar\Delta\hat{a}^\dagger\hat{a}$.

Check Points

- Derive the optomagnonic coupling Hamiltonian.

6.4 Coupled Equations of Motion and Fast Cavity Limit

The coupled Heisenberg equations of motion are obtained by using $\left[\hat{a}, \hat{a}^\dagger\right] = 1$, $\left[\hat{S}_i, \hat{S}_j\right] = i\epsilon_{ijk}\hat{S}_k$. We next focus on the classical limit, where we replace the operators by their expectation values:

$$\dot{a} = -i\left(GS_x - \Delta\right)a - \frac{\kappa}{2}\left(a - \alpha_{\max}\right)$$
$$\dot{\mathbf{S}} = \left(Ga^*a\,\mathbf{e}_x - \Omega\,\mathbf{e}_z\right) \times \mathbf{S} + \frac{\eta_G}{S}(\dot{\mathbf{S}} \times \mathbf{S})\,. \tag{6.29}$$

From Sect. 6.2, we know our optical cavity is an open system and the optical fields are subject to a decay rate. Here we introduced the cavity decay rate κ phenomenologically, its value is in general determined by the scpecific experimental setup. We also included the driving laser amplitude $\alpha_{\max}$ for the optical mode. This gives the steady state amplitude of the light field when it is not coupled to the magnetics and for zero detuning of the driving laser. We also added an intrinsic damping for the spin η_G, which can be due to phonons and defects and it is material dependent. This coefficient is denominated *Gilbert damping* and, whereas it does not change the magnitude of the spin vector, it causes a decay of the Larmor precession to the stable equilibrium of the spin. The equation of motion for the spin without coupling to the light reduces to

$$\dot{\mathbf{S}} = -\Omega\,\mathbf{e}_z \times \mathbf{S} + \frac{\eta_G}{S}(\dot{\mathbf{S}} \times \mathbf{S})\,,$$

which is known as the *Landau-Lifschitz-Gilbert equation*. We have encountered this equation before, abeit without the damping term.

We see hence that the light acts as a kind of effective magnetic field on the spin. Actually, since the field a depends on time, and the spin-light dynamics is coupled, retardation effects cause also dissipation for the spin, in a similar way in which an environment causes the dissipation term κ for the optical field. In the following we will obtain the effective equation of motion for the spin induced by the light, "integrating out" the light field. For that we have to resort to an approximation, which is denominated the *fast cavity limit*, where the dynamics of the light is much faster than that of the spin (sometimes it is also called the bad cavity limit, since it implies that κ is large). That means that the photons spend in average a very short time in the cavity, during which they "see" the spin almost as static.

The condition for the fast cavity limit to be valid is $G\dot{S}_x \ll \kappa^2$. In that case we can expand the field $a(t)$ in powers of $\dot{S}_x$. We write $a(t) = a_0(t) + a_1(t) + \ldots$, where the subscript indicates the order in $\dot{S}_x$. From the equation for $a(t)$, we find that a_0 fulfills the instantaneous equilibrium condition

$$a_0(t) = \frac{\kappa}{2}\alpha_{max}\frac{1}{\frac{\kappa}{2} - i\left(\Delta - GS_x(t)\right)}, \tag{6.30}$$

from which we obtain the correction a_1:

$$a_1(t) = -\frac{1}{\frac{\kappa}{2} - i\left(\Delta - GS_x\right)}\frac{\partial a_0}{\partial S_x}\dot{S}_x. \tag{6.31}$$

To derive the effective equation of motion for the spin, we replace $|a|^2 \approx |a_0|^2 + a_1^*a_0 + a_0^*a_1$ in Eq. 6.29 which leads to

$$\dot{\mathbf{S}} = \mathbf{B}_{eff} \times \mathbf{S} + \frac{\eta_{opt}}{S}(\dot{S}_x\,\mathbf{e}_x \times \mathbf{S}) + \frac{\eta_G}{S}(\dot{\mathbf{S}} \times \mathbf{S}). \tag{6.32}$$

Here $\mathbf{B}_{eff} = -\Omega\mathbf{e}_z + \mathbf{B}_{opt}$, where $\mathbf{B}_{opt}(S_x) = G|a_0|^2\mathbf{e}_x$ is the purely static contribution and acts as an optically induced magnetic field. The second term is due to retardation effects, and it reminiscent of Gilbert damping, albeit with spin-velocity component only along $\hat{\mathbf{x}}$ due to the chosen geometry. These are depicted in Fig. 6.2. Both the induced field $\mathbf{B}_{opt}$ and dissipation coefficient η_{opt} depend explicitly on the instantaneous value of $S_x(t)$:

$$\mathbf{B}_{opt} = \frac{G}{[(\frac{\kappa}{2})^2 + (\Delta - GS_x)^2]}\left(\frac{\kappa}{2}\alpha_{max}\right)^2\mathbf{e}_x \tag{6.33}$$

$$\eta_{opt} = -2G\kappa S\,|\mathbf{B}_{opt}|\frac{(\Delta - GS_x)}{[(\frac{\kappa}{2})^2 + (\Delta - GS_x)^2]^2}. \tag{6.34}$$

These fields are highly non-linear functions of the spin. Note that the optically induced dissipation can change sign! This leads to very interesting dynamics. Two distinct solutions can be found: generation of new stable fixed points (switching) and opto-magnonic limit cycles [16].

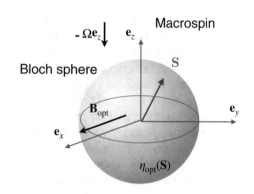

Fig. 6.2 Optically induced effective magnetic field $\mathbf{B}_{\mathrm{opt}}$ and dissipation coefficient η_{opt}. Together with the external applied magnetic field, which controls the Kittel mode frequency Ω, they govern the nonlinear dynamics of the macrospin $\mathbf{S}$ on the Bloch sphere

1. *Exercise: fill in the steps of the derivation above.*

Check Points

- What is the fast cavity limit?
- How do you obtain an effective equation of motion for the spin, and what is the meaning of each term?

6.5   Linearized Optomagnonic Hamiltonian

As we mentioned in Sect. 6.3, one can also treat the optomagnonic Hamiltonian in the limit of small oscillations of the spins, by using a truncated Holstein Primakoff expansion so that the spin ladder operators are replaced by linear bosonic operators. This allows to study the behavior of the system beyond the homogeneous magnetic Kittel mode studied in the previous two sections, but restricts the analysis to small displacements of the spins with respect of their equilibrium positions.

In order to proceed with the linearization, we consider spin wave excitations on top of a possibly nonuniform static ground state $\mathbf{m}_0(\mathbf{r})$,

$$\delta\mathbf{m}(\mathbf{r}, t) = \mathbf{m}(\mathbf{r}, t) - \mathbf{m}_0(\mathbf{r}). \tag{6.35}$$

For small deviations $|\delta\mathbf{m}| \ll 1$ we can express these in terms of harmonic oscillators, which correspond to the magnon modes. This is equivalent to a local Holstein Primakoff approximation. We can quantize the spin wave as

$$\delta\mathbf{m}(\mathbf{r}, t) \rightarrow \frac{1}{2}\sum_{\gamma}\left[\delta\mathbf{m}_{\gamma}(\mathbf{r})\hat{b}_{\gamma}e^{-i\omega_{\gamma}t} + \delta\mathbf{m}_{\gamma}^{*}(\mathbf{r})\hat{b}_{\gamma}^{\dagger}e^{i\omega_{\gamma}t}\right]. \tag{6.36}$$

In turn, the quantization of the optical fields can be written as

$$\mathbf{E}(\mathbf{r}, t) \rightarrow \sum_\beta \mathbf{E}_\beta(\mathbf{r}) \hat{a}_\beta e^{-i\omega_\beta t} \tag{6.37}$$

$$\mathbf{E}^*(\mathbf{r}, t) \rightarrow \sum_\beta \mathbf{E}_\beta^*(\mathbf{r}) \hat{a}_\beta^\dagger e^{i\omega_\beta t} \tag{6.38}$$

From Eq. 5.4.1 we obtain the coupling Hamiltonian linearized in the spin fluctuations [20]

$$\hat{H}_{MO} = \sum_{\alpha\beta\gamma} G_{\alpha\beta\gamma} \hat{a}_\alpha^\dagger \hat{a}_\beta \hat{b}_\gamma + \text{h.c.} \tag{6.39}$$

where

$$G_{\alpha\beta\gamma} = -i \frac{\theta_F \lambda_n}{4\pi} \frac{\varepsilon_0 \varepsilon}{2} \int d\mathbf{r} \, \delta\mathbf{m}_\gamma(\mathbf{r}) \cdot [\mathbf{E}^*(\mathbf{r}) \times \mathbf{E}_\beta(\mathbf{r})] \tag{6.40}$$

is the optomagnonic coupling. Note that $\hat{a}$ correspond to photon operators, while $\hat{b}$ correspond to magnonic ones. The Greek subindices indicate the respective magnon and photon modes which are coupled. The information on the specific shape and normalization of the magnon and optical modes is encoded in the respective mode functions $\delta\mathbf{m}_\gamma(\mathbf{r})$ and $\mathbf{E}_\alpha(\mathbf{r})$. Equations 6.39 and 6.40 allow to treat arbitrary geometries for the optical cavity, arbitrary magnetic ground states, and arbitrary spin wave modes.

The Hamiltonian in Eq. 6.39 is still nonlinear, since it involves products of three bosonic operators. This Hamiltonian therefore still contains "interacting" three-particle terms. In particular, it describes scattering processes in which a photon in mode β and a magnon in mode γ are annihilated creating a photon in the mode α, and the complementary process in which a photon α is annihilated creating a photon β and a magnon γ. To bring this Hamiltonian into a solvable, quadratic form, we linearize now the Hamiltonian Equation 6.39 in the optical fields. For that we consider fluctuations of the optical fields around a steady state solution $\langle \hat{a}_\alpha \rangle$, $\langle \hat{a}_\beta \rangle$

$$\hat{a}_\alpha = \langle \hat{a}_\alpha \rangle + \delta\hat{a}_\alpha$$
$$\hat{a}_\beta = \langle \hat{a}_\beta \rangle + \delta\hat{a}_\beta . \tag{6.41}$$

The steady state solutions $\langle \hat{a}_\alpha \rangle$, $\langle \hat{a}_\beta \rangle$ satisfy $\langle \dot{\hat{a}}_\alpha \rangle = 0$, $\langle \dot{\hat{a}}_\beta \rangle = 0$, where the time evolution is given by the coupled equations of motion dictated by the interaction Hamiltonian Eq. 6.39 plus driving and free terms, obtained by generalizing Eq. 6.28 to multiple modes. The average number of photons circulating in cavity mode α in steady state is simply given by $n_\alpha = |\langle \hat{a}_\alpha \rangle|^2$ and it is related to the input laser power. To linear order in the fluctuations defined by Eqs. 6.41, the Hamiltonian Eq. 6.39 reduces to

$$\hat{H}_{lin} = \sum_{\alpha\beta\gamma} G_{\alpha\beta\gamma} \left(\sqrt{n_\alpha} \delta\hat{a}_\beta \hat{b}_\gamma + \sqrt{n_\beta} \delta\hat{a}_\alpha^\dagger \hat{b}_\gamma \right) + \text{h.c.} , \tag{6.42}$$

which is a Hamiltonian linear both in magnon and photon operators. It contains two types of terms, denominated *parametric amplifier*, corresponding to those terms which simultaneously create or annihilate a photon and a magnon ($\delta\hat{a}_\beta\hat{b}_\gamma$ and $\hat{b}_\gamma^\dagger\delta\hat{a}_\beta^\dagger$), and *beam splitter*, which converts a photon into a magnon ($\hat{b}_\gamma^\dagger\delta\hat{a}_\alpha$), and vice-versa ($\delta\hat{a}_\alpha^\dagger\hat{b}_\gamma$). Which of the two types of processes dominates depends on which of them is in resonance, and can be tuned by the external laser driving. We note that the optomagnonic coupling constant $G_{\alpha\beta\gamma}$ is enhanced by the square root of the number of photons circulating in the corresponding cavity mode. This is similar to *optomechanics*, where light in an optical cavity couple to phonons [21].

6.6   Prospects in Cavity Optomagnonics

Cavity optomagnonic systems are the newest addition to a collection of platforms being studied nowadays with the aim of manipulating, processing, and storing quantum information. These systems, usually of nano and micro scale dimensions, are called *hybrid quantum systems* [22], since they combine different degrees of freedom (such as electronic, mechanical, photonic, or magnetic) to enhance functionality. For example, whereas optical photons are good carriers of information, they are not so good for information processing. Further examples of hybrid systems are nanoelectromechanical or optomechanical systems. A general underlying property of these is that they use collective excitations (such as phonons or magnons) whose properties can be engineered by proper design at the nanoscale.

A challenge in many of these platforms is that the coupling between the different degrees of freedom is weak, even taking into account the enhancement obtained by the use of a cavity. Strong coupling, together with low losses, are required for quantum information applications. This is because one should be able to process and transfer information before it is lost to the environment. In particular for magnonic systems, it has been shown that strong coupling to microwave photons is possible, by using a microwave cavity [23–26]. Note that in this case the coupling is resonant, meaning that the frequencies from both microwave and magnonic excitations can be matched, being both in the GHz range. The magnons in this case couple directly to the slow, oscillating magnetic field, as in ferromagnetic resonance experiments. We have not discussed this coupling in these notes, but it can be shown it has the form

$$\hat{S}^+\hat{a}^\dagger + \hat{S}^-\hat{a} \tag{6.43}$$

in terms of the spin ladder operators $\hat{S}^\pm$ and the microwave photons $\hat{a}$. This interaction converts a magnon into a photon and vice-versa. The optomagnonic coupling instead is parametric in the photon fields (coupling instead to terms of the form $\hat{a}^\dagger\hat{a}$). This is in general the case for non-resonant interactions, where the frequency mismatch has to be accounted for (note that optical photons have frequencies of hundreds of THz), and usually results in small intrinsic coupling values.

The fact that magnons can couple coherently both to microwaves and to light is however a big incentive to pursue the strong coupling regime also in the optical domain. That would allow coherent transfer of information from the microwave regime, where the information is usually processed (e.g. with superconducting qubits [27]), to the telecom regime, where information can be communicated through long distances and at room temperature with the help of optical fibers. We can expect that the next few years will bring many exciting advances in this field.

References

1. Jackson, J.D.: Classical Electrodynamics, 3rd edn. Wiley (1998)
2. Griffiths, D.J.: Introduction To Electrodynamics, 4th edn. Pearson (2014)
3. Nolting, W., Ramakanth, A.: Quantum Theory of Magnetism. Springer, Heidelberg (2009)
4. Stancil, D.D., Prabhakar, A.: Spin Waves: Theory and Applications. Springer, US (2009)
5. Castelnovo, C., Moessner, R., Sondhi, S.L.: Magnetic monopoles in spin ice. Nature **451**, 42 (2008)
6. Fennell, T., Deen, P.P., Wildes, A.R., Schmalzl, K., Prabhakaran, D., Boothroyd, A.T., Aldus, R.J., McMorrow, D.F., Bramwell, S.T.: Magnetic coulomb phase in the spin ice $Ho_2Ti_2O_7$. Science **326**, 415 (2009)
7. Morris, D.J.P., Tennant, D.A., Grigera, S.A., Klemke, B., Castelnovo, C., Moessner, R., Czternasty, C., Meissner, M., Rule, K.C., Hoffmann, J.-U., Kiefer, K., Gerischer, S., Slobinsky, D., Perry, R.S.: Dirac strings and magnetic monopoles in the spin ice $Dy_2Ti_2O_7$. Science **326**, 411 (2009)
8. Ashcroft, N.W., Mermin, N.: Solid State Physics. Cengage Learning Inc., New York (1976)
9. Auerbach, A.: Interacting Electrons and Quantum Magnetism. Graduate Texts in Contemporary Physics. Springer, New York (1994)
10. Lyons, D.H., Kaplan, T.A.: Method for determining ground-state spin configurations. Phys. Rev. **120**, 1580 (1960)
11. Kittel, C.: Introduction to Solid State Physics, 8th edn. Wiley, Hoboken, NJ (2004)
12. Landau, L.D., Pitaevskii, L.P., Lifshitz, E.M.: Electrodynamics of Continuous Media, 2nd edn. Butterworth-Heinemann (1984)
13. Cohen-Tannoudji, C., Dupont-Roc, J., Grynberg, G.: Photons and Atoms: Introduction to Quantum Electrodynamics, 1st edn. Wiley, Weinheim (1997)
14. Gerry, C., Knight, P.: Introductory Quantum Optics. Cambridge University Press, Cambridge, UK and New York (2004)
15. Meystre, P., Sargent, M.: Elements of Quantum Optics, 4th edn. Springer, Heidelberg (2007)
16. Kusminskiy, S.V., Tang, H.X., Marquardt, F.: Coupled spin-light dynamics in cavity optomagnonics. Phys. Rev. A **94**, 033821 (2016)
17. Osada, A., Hisatomi, R., Noguchi, A., Tabuchi, Y., Yamazaki, R., Usami, K., Sadgrove, M., Yalla, R., Nomura, M., Nakamura, Y.: Cavity optomagnonics with spin-orbit coupled photons. Phys. Rev. Lett. **116**, 223601 (2016)
18. Zhang, X., Zhu, N., Zou, C.-L., Tang, H.X.: Optomagnonic whispering gallery microresonators. Phys. Rev. Lett. **117**, 123605 (2016)
19. Haigh, J.A., Nunnenkamp, A., Ramsay, A.J., Ferguson, A.J.: Triple-resonant brillouin light scattering in magneto-optical cavities. Phys. Rev. Lett. **117**, 133602 (2016)

© The Author(s), under exclusive license to Springer Nature Switzerland AG 2019
S. Viola Kusminskiy, *Quantum Magnetism, Spin Waves, and Optical Cavities*,
SpringerBriefs in Physics, https://doi.org/10.1007/978-3-030-13345-0

20. Graf, J., Pfeifer, H., Marquardt, F., Kusminskiy, S.V.: Cavity optomagnonics with magnetic textures: coupling a magnetic vortex to light. Phys. Rev. B **98**, 241406(R) (2018)
21. Aspelmeyer, M., Kippenberg, T.J., Marquardt, F.: Cavity optomechanics. Rev. Mod. Phys. **86**, 1391–1452 (2014)
22. Kurizki, G., Bertet, P., Kubo, Y., Mølmer, K., Petrosyan, D., Rabl, P., Schmied-mayer, J.: Quantum technologies with hybrid systems. Proc. Natl. Acad. Sci. **115**, 3866 (2015)
23. Soykal, Ö.O., Flatté, M.E.: Strong field interactions between a nanomagnet and a photonic cavity. Phys. Rev. Lett. **104**, 077202 (2010)
24. Huebl, H., Zollitsch, C.W., Lotze, J., Hocke, F., Greifenstein, M., Marx, A., Gross, R., Goennenwein, S.T.B.: High cooperativity in coupled microwave resonator ferrimagnetic insulator hybrids. Phys. Rev. Lett. **111**, 127003 (2013)
25. Tabuchi, Y., Ishino, S., Ishikawa, T., Yamazaki, R., Usami, K., Nakamura, Y.: Hybridizing ferromagnetic magnons and microwave photons in the quantum limit. Phys. Rev. Lett. **113**, 083603 (2014)
26. Zhang, X., Zou, C.-L., Jiang, L., Tang, H.X.: Strongly coupled magnons and cavity microwave photons. Phys. Rev. Lett. **113**, 156401 (2014)
27. Tabuchi, Y., Ishino, S., Noguchi, A., Ishikawa, T., Yamazaki, R., Usami, K., Nakamura, Y.: Coherent coupling between a ferromagnetic magnon and a superconducting qubit. Science **349**(6246), 405–408 (2015)

Printed in the United States
By Bookmasters